Modern Aspects of Evolution

A P Brookfield

Hutchinson

London Melbourne Sydney Auckland Johannesburg

Hutchinson & Co. (Publishers) Ltd

An imprint of the Century Hutchinson Publishing Group

Brookmount House, 62–65 Chandos Place, Covent Garden, London WC2N 4NW

Hutchinson Group (Australia) Pty Ltd
16–22 Church Street, Hawthorn, Melbourne, Victoria 3122

Hutchinson Group (NZ) Ltd
32–34 View Road, PO Box 40–086, Glenfield, Auckland 10

Hutchinson Group (SA) (Pty) Ltd
PO Box 337, Bergvlei 2012, South Africa

First published 1986 by Hutchinson and Co. (Publishers) Ltd
© A P Brookfield 1986
Illustrations © Hutchinson and Co. (Publishers) Ltd 1986

Typeset in 11 on 13pt VIP Plantin by
D P Media Limited, Hitchin, Hertfordshire

Printed and bound in Great Britain by
Anchor Brendon Ltd, Tiptree, Essex

Illustrations by the author

British Library Cataloguing in Publication Data

Brookfield, A. P.
  Modern aspects of evolution
  1. Evolution
  I. Title
  575   QH366.2

ISBN 0 09 164571 9

Contents

Preface   iv

1 Some ideas of the origin and history of life on earth   1

Spontaneous generation; life from space; special creation;
steady state theory; evolution; summary

2 Evidence that evolution may have occurred   19

The fossil record; biogeography; comparative anatomy;
biochemistry and physiology; observed evolution;
summary

3 The raw material for evolution   76

Mendel's work; gene interaction; genetic code; analysis of
proteins; mutation – types and rates; control systems of
gene action in prokaryotes and eukaryotes; summary

4 The genetic structure of populations   109

Characteristics of populations; gene frequencies in an
idealized population; summary

5 The mechanism of evolution   125

Changes in meiosis; mutation; an open population –
migration pressure and gene flow; population size and
non-random mating; selection; summary

6 The results of evolution   175

The species concept; methods of classification; types of
isolation; types of speciation; stages of speciation; gradualism and punctualism; the concept of adaptation; future
developments; summary

Appendix A   The application of the Hardy-Weinberg law   209
Appendix B   Experiments in selection   212

Glossary 214   *Bibliography* 217   *Index* 219

Preface

Anyone who is interested in biology must be interested in the ways in which present day organisms have been derived from those which existed in the past. Many biologists feel that the incredibly diverse and complicated structures and variation in function shown in living organisms can only make sense if considered within an evolutionary framework – a framework which explains the descent of present day organisms from their very different ancestors.

Biologists may argue about the ways in which such changes could have been produced during the course of evolution and the arguments often 'spill-over', through the media of television, radio and newspapers, to a general audience. There has probably been more controversy amongst the general public about evolution, starting with the great arguments after the publication of Darwin's *The Origin of Species*, than about any other scientific topic. Evolutionary theory continues to branch out in different directions to take account of the new information which is constantly being provided by biological research in various fields, and it is this fact which makes the whole subject so fascinating. This book summarizes many of the past and present sources of controversy so that the reader may begin to judge the evidence, and the wide list of references could suggest further reading on the varied aspects covered.

I have been very fortunate in the assistance I have received while writing this book and for which I am very grateful. My younger son, Dr John Brookfield, has been generous with his time in reading the typescript at various stages and most helpful in his comments. My elder son, Dr David Brookfield, devised the computer program and has also helped in many other ways. Dr Michael Roberts has also suggested various improvements to the manuscript, which I have been glad to incorporate. I much appreciate the help and encouragement given by Pat Rowlinson at Hutchinson Education. I would also like to thank my students who, by their often astute questioning, stimulated me to try to organize the material into a form which is accessible to them and which I hope they will enjoy reading.

A P Brookfield

1

Some ideas of the origin and history of life on earth

One of the delights in being interested in the natural world is the enormous number of different organisms present in it. More than one and a half million **species** of plants, animals and micro-organisms have been given names and there are likely to be many others which have not yet been discovered. We all have our own particular interests. Some people can spend a lifetime studying beetles, while others find excitement in seeing a bird which is new to an area or go hunting for rare plants. The earth is teeming with life. Even a small garden will contain a vast number of different species of organisms. You know how quickly life will take over an originally bare region. A small amount of water left in the bottom of an old bucket in a garden will become inhabited by micro-organisms carried there by the wind or on other organisms. Those of you who are gardeners will know how rapidly a carefully tended area of soil will become weed infested if you have a few days holiday.

Where have all these different organisms on earth come from? The short answer, of course, and the one which might be most frequently given, is that they came from their parents (or parent if they were asexually reproduced). But this only pushes the question one stage further back. The parents came from the grandparents and so we would need to think back, stage by stage, to the beginning of life on earth. Has it always been accepted that organisms could only be formed by some sort of reproduction in other organisms? Could they be derived from non-living matter by a process of spontaneous generation? If life did arise from non-living matter, when and where did this happen? Was it a once and for all occurrence, or did it happen many times in different parts of the world? Did life originate on earth at all, or are all living organisms on earth the descendants of living matter which came from space? Have the organisms which we see on earth today come down unchanged through a long history of life, having been created in their present form at some particular and knowable (if not yet known) time? Are present day organisms different from those which were on earth millions of years ago? Certainly there is evidence from fossils that some plants and animals no longer exist. Are these present day organisms the result of a long and continuing series of changes which have produced the large diversity of species which we now see? If such changes, which we may call **evolution**, have occurred what possible mechanisms could have brought them about?

As you see, there are many, many questions which can be asked about the origin and derivation of life on earth. The questions are not new. They have been asked at many different times in the past and the answers which were accepted as satisfactory depended on the state of knowledge about the natural world at the time. Let us take a further look at the questions and consider what answers to them have been given.

SPONTANEOUS GENERATION

If life on earth was not the result of a special creation at one time, or the result of an invasion of simple life forms from space, then one must assume that life arose on this planet from non-living material. However, such an origin of life through a series of chemical reactions would not be described as spontaneous generation. We return to consider suggestions about the origin of the molecules, such as the enzymes and the **self-replicating nucleic acids**, which are essential for life in a later chapter. The idea of spontaneous generation assumes that complex living organisms such as insects or worms could suddenly be produced from dust, mud or water. The idea has been around for a long time. **Aristotle** (384–322 BC) wrote about a large number of carefully observed natural phenomena including sleep in animals, the life history of the cuckoo, the differences between males and females in many species, and the reasons for animal migration. He discussed how animal reproduction occurred and, although he thought that some animals mated and produced offspring, he thought that other animals arose spontaneously from water or mud.

Spontaneous generation implies a continuing series of creative acts in which organisms could suddenly arise from non-living matter or just as suddenly arise from organisms which were very different from themselves. The idea arose mainly from a lack of information about the life histories of organisms. Some animals have immature stages which are very different from the adult. A sudden appearance of adult mayflies, for example, flying above a pond might not have been seen to be related to a disappearance of mayfly nymphs within the pond water. The maggots which appeared on dead carcasses could have seemed to have developed from the decaying flesh. Even if blowflies (bluebottles) had been seen on the dead animal there was no obvious reason to relate the crawling, white maggots with the flying, shiny, metallic coloured adults. The close physical resemblance between some animals and other objects was also a source of confusion. The hairworms (phylum Nematomorpha) are extremely long threadlike worms with no obvious head which live in fresh water and soil and it was formerly thought that they arose from any hairs from a horse's tail which happened to fall into the pond or onto the soil.

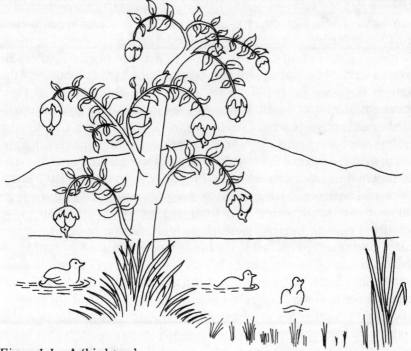

Figure 1.1 A 'bird tree'

Centuries ago it was thought that birds such as ducks developed from the buds of special 'bird trees'.

Mice, which had been attracted from the surrounding area to a grain store, could suddenly appear in large numbers and it seemed sensible to assume that they had developed from the wheat kernels. Sometimes confusion arose because the animals were migratory. Large obvious birds like ducks and geese could suddenly be found swimming on lakes where there had been none a few days before. There were many stories to explain their appearance, such as a suggestion that the birds had developed from the buds of trees which overhung the lake.

The stalked barnacles found on the rotting timbers of old ships were thought, in the far North, to develop into barnacle geese. These barnacles were the subject of a long investigation by Frederick the Second of Sicily in the Thirteenth Century. He sent envoys to the North to bring back barnacle specimens for study, so that he could see if they would turn into geese if given the right conditions. The barnacles were found to have no features of birds and it was concluded that the birds were migrants which suddenly appeared from other regions.

The idea of the spontaneous generation of such large animals as ducks, geese and mice was gradually discounted as more was learned of their ways of life. Experiments carried out by Francesco Redi in 1668 provided evidence that maggots did not develop from decaying meat. He placed pieces of meat in two jars, the open end of one of which he covered with cloth. Each piece of meat was open to the air but only the meat in the open jar could be visited by flies. Both pieces of meat putrefied; the meat in the open jar also became maggot infested. Further experiments were carried out in the Eighteenth Century by **Spallanzani**, and **Louis Pasteur** in the middle of the Nineteenth Century provided evidence that micro-organisms were reproduced by existing parental micro-organisms and did not arise spontaneously. Broth which had been sterilized by heat and kept in sealed flasks did not go bad whereas similar broth, sterilized and left open to the air, soon supported enough micro-organisms for putrefaction to occur.

It is now accepted that all living organisms are derived by reproduction from other living organisms. Of course, this still requires some explanation, firstly for the origin of the molecules needed for life, and secondly for the way in which these molecules collected together to make the first cells. We return in a later chapter to consider how this might have happened on earth; an alternative suggestion is that life is not only found on earth but is widespread throughout the universe.

LIFE FROM SPACE

All the elements present on earth had their origins in the nuclear reactions which took place on stars and from this point of view everything on earth had a cosmic origin. It has been suggested that life could have been propagated from one solar system to another by the spores of micro-organisms. The idea was popular in the Nineteenth Century as the theory of **panspermia**. Recently **Crick** (of DNA fame) and **Orgel** have suggested a process, which they have named directed panspermia, by which sterile planets such as early earth were 'seeded' with life from other solar systems. The 'seeding' was done by intelligent beings whose stage of evolution was billions of years ahead of our own. **Hoyle** and **Wickramasinghe** also believe that life reached earth from space.

SPECIAL CREATION

Special creation suggests that all the organisms which are present on earth were created by some supernatural being. The type, or types, of supernatural being which one accepts depends, of course, on the culture in which one is reared. There are many creation stories in the different religions of the world. The Shinto religion in Japan claims that the Emperor is the descendant of the sun-goddess Amaterasu, and that the rest of the Japanese people are the descendants of lesser gods. The islands of Japan are seen as having been born from the womb of the first woman, Izanami. Members of one of the tribes of North American indians, the Navaho, believe that human life began in the fourth of five worlds, the Fifth World being the present day world. Life began in the First World in the form of twelve kinds of insect people living in a dark land. They survived until the Fourth World when, under instruction from the gods, they laid a perfect ear of white corn and a perfect ear of yellow corn on eagle feathers between sacred buckskins. The life-giving White Wind blowing from the East and the Yellow Wind from the West then caused the first man and the first woman to come to life from the corn. After some time in the Fourth World, these first human beings then ascended to the present Fifth World through a small hole in the sky.

When the term 'special creation' is used in Britain or in the Protestant areas of the United States of America, we tend to think of the Jewish-Christian idea as described in Genesis of the Old Testament. You must remember that although the people who hold this view might dismiss non-biblical accounts as creation myths, devout members of other religions must regard Genesis as a myth in the same fashion.

Let me remind you of the creation account in the first chapter of Genesis. The creation of the earth and all the organisms which live there took six days. The first day was needed to make light and darkness. The second saw the creation of the heaven and the earth. On the third day dry land was separated from the oceans, and the land was clothed with plants. The sun, moon and stars were made on the fourth day. All the animals which moved in the seas and those which flew in the air were created on the fifth day; those which lived on land were made on the sixth day. Man was also created on the sixth day to 'have dominion over the fish of the sea, and over the fowl of the air, and over every living thing which moveth on the earth'.

So in what Genesis 1 describes as six days (although the compilers of the chapter might not have meant six of our normal, twenty-four hour days) all the organisms were created on earth. But that isn't the whole story. If you read on into chapter 2 you will realize that, whereas man is the culmination of the creation process as described in chapter 1, he is the first organism to be created according to chapter 2, verses 7–19. Already there are problems in the literal acceptance of the Genesis account. Read on further through the description of Adam and Eve and their fall from grace, and through the account of their descendants, and you will come to the description of Noah and the ark and the great flood in chapters 6–10. The great catastrophe of the flood killed all those organisms which had previously been present on earth, other than those which had been taken into the ark with Noah and his wife, and their sons and their wives. So all present day organisms must be the descendants of those which survived the catastrophe, and not necessarily those which had been created earlier during the six days of creation.

What are we to make of the claims of those who hold this account of the origin and derivation of life on earth to be literal truth? A Gallup poll held in the United States of America in 1979 found that about half of the people questioned believed that God created Adam and Eve to start the human race, and so the acceptance of the Genesis account is widespread. Many accept the account because they have never considered alternatives. Present day creationist ideas depend on a fundamentalist belief that the Bible is true in all its details. However, the majority of Christian biblical scholars would not support such a view. Many Nineteenth Century creationists were not so restricted by fundamentalism.

They tried to find out as much as they could about nature and natural processes hoping that this would show the workings of God's mind. Geologists, such as Louis Agassiz (1807–1873), did not study rock formations in order to find evidence of the biblical flood, but to try to understand the processes which had been involved in laying down the rocks. His investigations led him to the conclusion that there had been many catastrophic events in the history of the earth which could have affected the organisms present at the time. He proposed in 1840, for example, that large areas of the northern continents had been covered by ice sheets in the geologically recent past. Scientists like Agassiz found no difficulty in holding the conclusions reached as a result of many scientific investigations, and a belief in the creation of organisms by God. Over the past two decades there has been an attempt to explain what is described in Genesis by using the language and methods of science, rather than suggesting an acceptance because of faith and the modern theory of scientific creationism has been developed.

But what do we mean by 'the language and methods of science'? As biologists, you must be familiar with the idea of a scientific theory. Some theories bear the names of their originators, such as **Darwin's** theory of evolution by natural selection, or **Lamarck's** theory of the inheritance of acquired characters. A scientific theory is basically an intelligent guess to account for certain observed facts. The theory may then be used to make predictions which may be tested by further observation or experiment. The information gathered may be sufficiently similar to the prediction for it to support the theory. If the new information differs significantly from the prediction, then an explanation for the difference must be sought. The theory itself must be altered to take account of the new information if a satisfactory explanation cannot be found. This means, of course, that a theory cannot be *proved* to be correct. Even if it is supported by all the available data, there is always a chance that some future experimental result or observation might not fit. Further experiments or more observation would then be needed to decide if the theory should be altered, and how this should be done. Thousands of experiments, the results of which support a theory, do not create as much excitement as one which produces the result which appears to refute a long-established, well-accepted theory. Sometimes this one vital experiment turns out to give different results when repeated by other scientists. The experimental design must then be scrutinized to see if the first result may be considered to be valid. The whole business of science is this constant testing and retesting of theories in the light of new evidence. At any time many theories may be seen to be 'under attack' but this is how science proceeds.

Scientists can share the knowledge gained from experiments carried out in laboratories all over the world because, although they may speak and write in different languages, the 'language' of their scientific method is universal. Of course the acceptance of results which come from experiments in different geographical areas or at different times depends not only on an agreement on experimental methods but also on an assumption that matter behaves according to universal laws, and does not change its characteristics and behaviour in unpredictable and capricious ways. Scientific thinking assumes that everything has a cause, and that causes and effects can be studied because they behave according to natural laws.

This is one of the areas in which it becomes difficult to equate the ideas of creationism with scientific thinking. The introduction of supernatural causes by means of a Creator removes the problem being studied from the realm of science because such supernatural causes would, by their very nature, be unpredictable and untestable. If the account of the creation of life as given in Genesis, or in any other account given in other cultures, is regarded as God-given and therefore unchallengeable, it cannot be discussed in scientific terms. As we have said, the whole business of science is the constant testing and challenging of theories and their alteration if the evidence does not fit. It is the theory which is altered and not the evidence. Modern scientific creationists, although they may appear (and claim) to work scientifically, must always be surrounded by the constraints of trying to fit the evidence to a rigid and permanent story. They see the theory of evolution as being unprovable and extend this into an assumption that this means it must necessarily be incorrect. They claim that the only possible alternative is special creation. Controversy amongst biologists about the various mechanisms by which evolution might have occurred is seen as negative evidence for evolution and therefore positive evidence for special creation.

Problems which could arise for creationists from the Genesis account, such as the excessive age of the earth as suggested by **radio-isotope dating**, are regarded as irrelevant in the context of an all-powerful God who could, as he did in the flood, alter any aspect of creation or the state of the earth as he wished. Modern creationists look for evidence of the flood in the rocks and dismiss the claims for the rock strata to be evidence of hundreds of millions of years of sedimentation because the flood evidence is not there. They ask why man, *if* his history is millions of years old, has developed agriculture only in the past ten thousand years. According to Genesis, Adam and Eve had these skills from the very beginning. They work out the dimensions of the ark from the biblical accounts and then discuss the logistics of taking in and supporting all the known species of animals (the fish, fortunately, could remain outside swimming about in the flood water). Dinosaurs, of course, could have been a problem because of the great size of many of the species, although it has been suggested that baby dinosaurs were taken into the ark for this reason. The solutions to the problems of feeding all these animals and, probably more important, of dealing with their faeces and excretory products are not given.

These may be regarded as flippant arguments against the fundamentalism which is now called scientific creationism. What is important, however, is the way in which evidence may be selected and distorted to fit an accepted story. Of course we all tend to pay more attention to evidence which supports our favourite theory. Most people find it difficult at first to accept that some evidence may point in another direction, and that they may need to change their view on a topic. But that is one of the consequences of reading more, and learning more, about anything. You would not expect to believe at the age of eighteen all the ideas which might have seemed to be obviously true when you were seven years old. We must not behave like blinkered horses with their very limited field of view, but try to see and consider as much of the range of evidence as possible.

So far we have mentioned the possibilities that life started, in a simple form, on earth or that it started in some other region of the universe. Both views have their supporters and both can be investigated by scientific means even though no one can go back in time to see what really happened. But both theories are concerned with the origin of life and not with its subsequent history on earth. We must now consider whether living organisms have changed during their history in a process of evolution or whether there is some other explanation for the diversity of species. Such an alternative to evolution is the steady state theory.

STEADY STATE THEORY

This proposes that species have existed throughout time and did not have a beginning. Although it accepts that certain species have become extinct, it claims that no species has changed into another species. One of the problems found in trying to interpret the fossil record is that it is often impossible to find a series of fossils which show transitional forms. The transition could show how a species either changed gradually over a long period of time into a different species, or produced the branch line which eventually became a different species. It was thought that the gaps in the record would be filled as further investigations were made. However, if species do not change into other species there can be no transitional forms. So the gaps in the fossil record could be seen as evidence for the steady state theory. They could also, of course, support special creation if this is taken to mean that each species was separately created and then remained fixed (but not the fundamentalist biblical view of special creation).

Another proposal to account for the gaps in the fossil record lies very firmly within an evolutionary framework.

EVOLUTION

By evolution, we mean that the organisms found on earth at the present time are the changed descendants of earlier organisms. The methods by which these changes could have been produced have been, and are at present, the source of much controversy.

Noted theorists

Evolutionary ideas have a long history. **Anaximander**, who lived in Greece from 611 to 547 BC, argued that life started in the sea and that organisms could gradually change from one form to another. However, the biblical ideas of special creation and the unchanging nature of species became accepted and it was not until the Eighteenth Century that scientists again began to think in evolutionary terms. **Erasmus Darwin** (the grandfather of Charles Darwin) believed in the evolution of organic life to its culmination in man, and thought that there could be variation over a long period of time by 'artificial or accidental cultivation' such as occurred in domestic animals which had been bred for certain characteristics. He also suggested that useless organs could be lost over a similarly long period of time. It seemed to him that many male animals acquired exclusive possession of females by means of weapons such as tusks or horns, and that this was accomplished best by the strongest males. So he thought that 'the strongest and most active animals would propagate the species which would thence become improved'.

However, the ideas involved in evolution were held by few scientists even by the early years of the Nineteenth Century. In 1809 **Jean Baptiste de Lamarck** published his theory of evolution in which he reached two main conclusions:

1 there is a progression from the simpler to the more complex kinds of organisms;
2 organisms are capable of changing their form during their lifetime in response to changes in the environment and these changes (acquired characters) may be passed on to their offspring.

Organisms can, and do, change their form in response to environmental changes – plants growing in unilateral or non-uniform light show the unequal growth patterns which are characteristic of phototropism, and most organisms which are short of food show stunted growth. However, such changes would have to be passed on to their offspring to support Lamarck's **theory of the inheritance of acquired characters**. Lamarck was unaware of the nature of genetical systems but his theory not only stated his belief in evolution but suggested a method by which evolutionary change could have come about.

Other scientists of the early Nineteenth Century also thought that the environment might have direct effects on organisms and that any changes which resulted could be passed on to their offspring. However, even then not all scientists who believed in evolution agreed as to the mechanisms which could have brought it about.

The two men who were to prove to be the most important in the field of evolution in the century were **Charles Darwin** and **Alfred Russell Wallace**. Darwin was born in 1809 and Wallace in 1823. Their lives were very different in many ways. Darwin was part of a wealthy family, whereas Wallace always had to earn his own living. Darwin had graduated from Cambridge University after a time at Edinburgh University. Wallace had been taught land surveying by his brother but his biological knowledge had been acquired by reading and observation. What they had in common was a great amount of travel to various parts of the world. In Darwin's case this occurred in the five year period between 1832 and 1837 in which he had sailed round the world in HMS *Beagle* as a companion to Captain FitzRoy. Wallace had gone to the Amazon with his friend H W Bates in 1848 and spent some time collecting specimens for Kew Gardens. He later returned to England (suffering shipwreck on the way) then set off for the Far East to collect more specimens. The journeys made by the two men can be seen in figures 1.2 and 1.3.

Both Darwin and Wallace had read an important book which had been published in 1798. This was *An Essay on the Principle of Population* by Thomas Malthus. Malthus was the first author to describe, in quantitative terms, the effect of unchecked population growth on the amount of available food. Darwin, having returned from his *Beagle* voyage, had married and continued to work on the notes he had made about the different species he had seen during the voyage. Wallace continued journeying around the Far East, very often ill and in difficult circumstances. Although the two men corresponded with each other as a result of a paper by Wallace published in *The Annals and Magazine of Natural History* in 1855, each was working independently on a theory that might explain how evolutionary change could have been produced. In 1858 Wallace sent off his conclusions to Darwin, and it was decided that this paper and a paper by Darwin which reached approximately the same conclusions should be read at the next meeting of the Linnaean Society of London.

Wallace's paper put forward two propositions.

1 The population size of a particular species remains, in general, constant. Its potential increase is checked by external factors, particularly the food supply.
2 The comparative abundance or scarcity of species is entirely due to their efficiency at fighting these external factors. As more animals are born than can be supported by the food supply, there must be an elimination of the less well adapted. The most healthy would escape the effects of disease; the strongest, swiftest or most cunning would escape from enemies; the best hunters could escape from famine. Any variation from the average form would make the animal more or less capable of surviving.

Darwin's paper also suggested that the potential increase in population size due to the great **fecundity** of organisms is kept in check by 'recurrent struggles against other species or against external nature'. The 'struggles' with other species were not physical but in the sense of competition for the limited resources of a harsh environment.

The propositions put forward in the two papers were extended in 1859 when Darwin's book *On the Origin of Species by Natural Selection* was published. Darwin's argument was based on two observable facts:

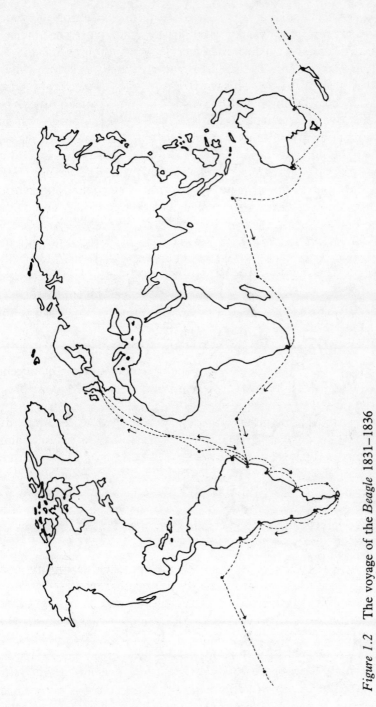

Figure 1.2 The voyage of the *Beagle* 1831–1836

Two years between July 1832 and July 1834 were spent in surveying the coast of South America between Montevideo on the east and Valparaiso on the west.

Figure 1.3 Some of Wallace's Far Eastern journeys 1854–1862

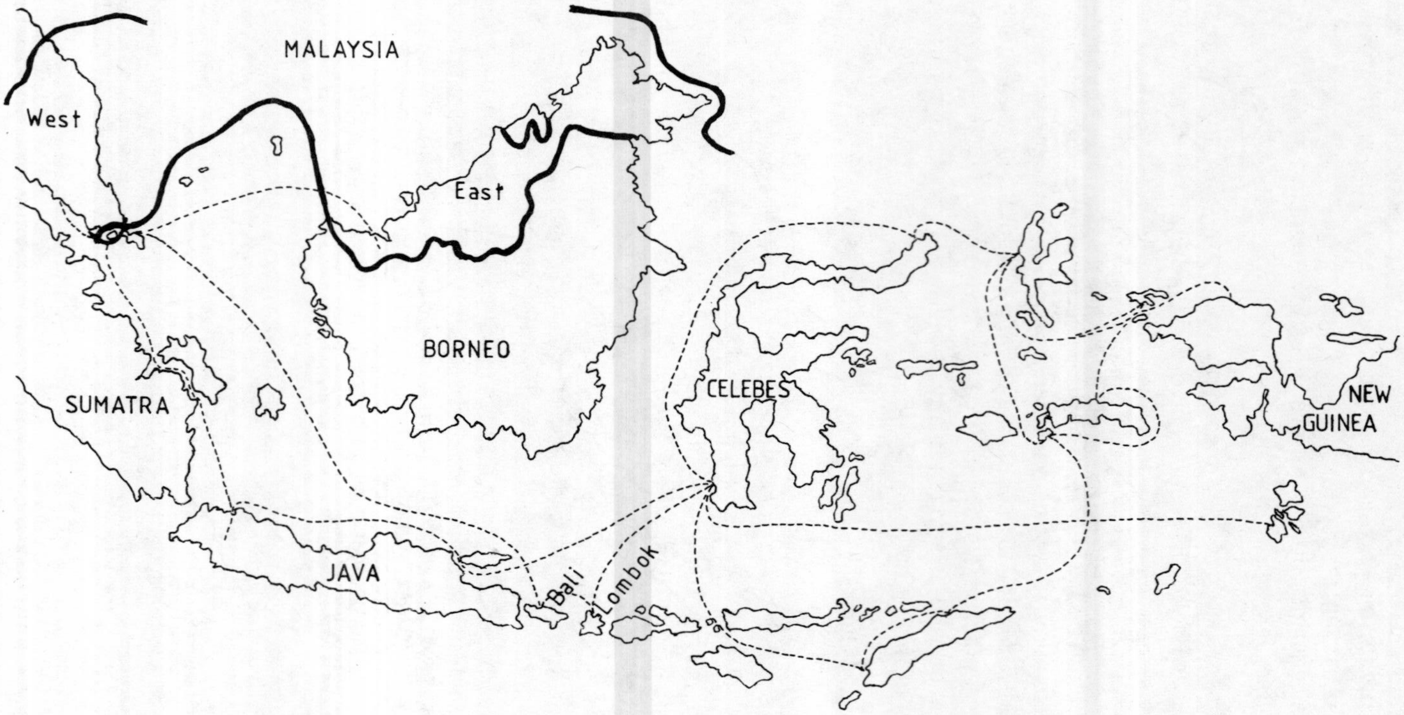

1 organisms vary, and many of these variations are inherited by their offspring;
2 organisms produce more offspring than can possibly survive as population sizes remain approximately constant.

From these two facts he concluded that those offspring which varied most strongly in directions favoured by their local environment would survive best to produce their own offspring. This being so, it is possible that the characters of the population could gradually change in such a direction by means of this **natural selection**. Darwin, writing as he was when Mendel's work on inheritance had not been published, found great difficulty in accounting for the production and maintenance of inheritable variation. His broad scheme required that such variations must be random and not in any 'preferred' direction if natural selection was to have a creative role in determining the direction of evolution. The variation must occur in fairly small steps. There must also be a substantial amount of inheritable variation present in a population if the population is going to be able to survive through changes in the environment, whether they occur in space or time.

The idea that some members of a population might survive longer than others, giving them an increased chance of reproducing, was later summed up by **Herbert Spencer** in the phrase **'the survival of the fittest'**. It was taken literally by many to mean that the larger and stronger animals would survive by fighting the smaller and less aggressive. But what matters is not just survival but also reproduction. The strongest or largest variety will not become commoner if it is sterile, or if its reproduction rate is much lower than that of other members of the population. So 'fitness' is a measure of relative adaptation (which allows some organisms to survive better in a particular environment) and also of reproductive success.

Neo-Darwinism

The theory of evolution by natural selection was accepted by many scientists of the late Nineteenth Century, even though they did not know how the genetic variation could be produced and maintained. However, the rediscovery of Mendel's work in 1900 provided just what was needed to put the theory on a firmer base. The following thirty years of work in the new, exciting field of genetics showed that small genetic mutations occur with great frequency and with no preferred direction to provide copious inheritable variation. Darwinism and genetics were merged to make a new 'synthetic' theory of evolution by natural selection, which was named **Neo-Darwinism**. Books such as *The Genetical Theory of Natural Selection* by R A Fisher in 1930 and *The Causes of Evolution* by J B S Haldane in 1932, and the important paper *Evolution in Mendelian Populations* by Sewall Wright in 1931 amplified Darwin's theory in the light of the chromosome theory of inheritance and the new ideas of population genetics.

Neo-Darwinism has been considered, for much of the time since the early 1930s, to be the best available theory to account for evolutionary change. However, recent research in genetics and in other fields has caused the theory to be re-examined. Discussions over the possible mechanisms by which evolution could have occurred are again producing excitement. Very many biologists believe that the factual information which is available supports the theory of evolution by natural selection. Other biologists, while accepting that evolution occurs and that populations change in their genetic constitution through time, argue that the changes result from chance events and so are random. The two views are not incompatible. Chance events which may eliminate some members of a population, irrespective of their adaptations, do not necessarily mean that those which are left cannot be affected by natural selection. It may be that chance events are very important in particular circumstances, and could produce large and relatively rapid changes in the population. The variations on which natural selection could work may result from small or large genetic changes. The rate at which a population might change as a result of natural selection is likely to be related to the rate at which the environment changes, and this may be very variable.

An acceptable theory of evolution must be able to encompass all the evidence from various fields of study. An acceptance, in the light of observational and experimental evidence, that natural selection is important in producing changes in populations does not refute other explanations which may be more applicable in certain circumstances.

There are many aspects of evolutionary theory which are now in the melting pot. One of the problems is that many people regard the terms evolution and Darwinism as being synonymous, and think that the first necessarily implies the second. But changes may occur in the genetic constitution of populations which result neither from natural selection nor from the chance external events we have mentioned. Sometimes mutations occur which seem to produce no phenotypic effects and which cannot increase or decrease the organism's adaptation to the environment. We return to consider such **neutral mutations** in chapter 5. The Lamarckian idea of the inheritance of acquired characters has recently been spotlighted by Steele's work in immunology, in which he claimed that immunity acquired by a mother mouse can be inherited by an offspring. Although his results have not been replicated by other eminent investigators in the same field and the experimental methods have been, and are being, reassessed, great interest was produced by the publication of his initial findings.

In this chapter we have suggested some of the views which are, or have been, held to account for the origin and subsequent history of life on earth. Some ideas, such as the spontaneous generation of complex organisms either from non-living matter or from other organisms which are very different, are not now accepted. The concept of special creation, if it is taken to involve the fitting of evidence to an unchallengeable statement of what was brought about by supernatural means, is not discussable in scientific terms. Whether organisms remain fixed in their species for ever, or whether species are derived from other species during a long process of evolution may be argued in the light of the evidence which is available. The author of this book takes the latter view and hopes that the remaining chapters of the book, in which the evolutionary standpoint is taken, may help you to understand the present controversies so that you will be in a better position to judge for yourselves.

SUMMARY

There have been, and still are, many arguments about how **life started** on this earth, and about what has happened to organisms **since then**. Some theories suggest that life did not start here at all but that **primitive organisms** came, by various means, to earth **from space**. In the past it has been thought that **more advanced organisms** might have been produced from non-living matter by a process of **spontaneous generation**. People who held this view also believed that organisms could suddenly arise from very different organisms. **Creation stories**, which describe how all the different organisms were created, are found in the holy books and traditions of many different religions. However, as these all depend on **supernatural events** which cannot follow the **natural laws** on which all scientists depend, they cannot be considered to be part of any science or be studied by scientific means. It is possible, however, by scientific means to try to find out how life might have started, in the form of **simple self-replicating molecules**, on earth and this topic is discussed more fully in chapter 2.

If life started here in such a way or came as primitive organisms from space then obviously there must have been a **long process of change** to produce all the different forms which are present today or which we know have existed although they are now extinct. Such a process of change is called **evolution**.

FURTHER READING

Gould, Stephen Jay, 'Evolution, theology and the Victorian scientist', *Nature*, vol. 285, no. 5763 (29 May 1980), p. 343.

Harper, G H, 'Alternatives to evolutionism', *School Science Review*, vol. 61, no. 214 (1979). (Also see replies in vols. 61 and 62.)

Hoyle, Fred, Wickramasinghe, Chandra, *Evolution from Space* (Dent, 1981).

Hoyle, Fred, Wickramasinghe, Chandra, *Why Neo-Darwinism Does Not Work* (University College of Cardiff Press, 1982).

Ruse, Michael, 'Darwin's theory – an exercise in science', *New Scientist*, vol. 90, no. 1259 (25 June 1981).

Steele, E J, *Somatic Selection and Adaptive Evolution. On the Inheritance of Acquired Characters* (Williams and Wallace, Toronto/Croom Helm, London, 1980).

Tudge, Colin, 'Lamarck lives – in the immune system', *New Scientist*, vol. 89, no. 1241 (19 February 1981).

Williams-Ellis, Amabel, *Darwin's Moon – A Biography of Alfred Russell Wallace* (Blackie, 1966).

2

Evidence that
evolution may have occurred

There is a large amount of evidence which may support the view that populations of organisms change through time. The idea of change through time involves the problem that we are here now, and so were not around to see what was happening in the distant past. We must rely on evidence of various types to tell us what might have happened over hundreds of millions of years. Some of this evidence lies in a **fossil record** which, as fossilization is such a chance affair, is likely to be imperfect. However, a fossil can provide information not just of the species of organism, but also of the environmental conditions in that area at the time. Present day monkeys, for example, are known to live in hot regions on earth. If fossils from monkeys are found in cold areas then two possible conclusions may be drawn. It is possible that the monkeys which were fossilized could have been adapted to life in a cold climate. Alternatively, it is possible that those areas were very much hotter in the past. If fossils of other animals and plants which now live in hot regions were found with the monkey fossils, it is more likely that the second conclusion is correct.

Other evidence may be found in **biogeography**, the study of the distribution of organisms in various parts of the earth. Organisms live in conditions to which they are adapted. However, similar environmental conditions may be found in different areas of the earth, but the organisms living in them may be very different. On the other hand, rather similar organisms may be found in geographically very separate areas. There must be some explanation for what may seem, at first, to be a very strange distribution.

Biogeography studies living organisms and tries to account for their present day distribution on earth by considering what might have happened to them in the past. In a similar way present day organisms may be compared with each other **anatomically and physiologically** to see if any close relationships may be found. The relationships may suggest a pattern of descent from ancestors which they had in common, making it possible to draw an evolutionary tree.

Sometimes the genetic changes in populations which form the basis of evolution may be seen to act in time scales which are observable by investigators, and so this evidence is more direct. Most of the **observed changes** are within species and therefore do not really help us to decide how new species could have arisen. New species, however, have been produced artificially in ways which are also very likely to have occurred naturally and this type of investigation can help us to understand what might have occurred at many times in the past.

Let us now consider each of these broad categories of investigations to see if the evidence they produce can support the view that evolution has occurred, and is occurring, on earth.

THE FOSSIL RECORD

Fossilization

The study of palaeontology involves not only the identification of fossils but also the study of the environmental conditions which prevailed when the fossilized organisms were alive. Fossils have been called 'the window into the past', but one of the problems involved in using them is that fossilization is a very hit-and-miss affair. The vast majority of organisms which have lived on earth have left no tangible evidence of their existence. Very occasionally the bones or other hard parts of an organism like the arthropod cuticle may be left undecayed after the rest of the organism has rotted, to be covered subsequently by layers of vegetation or other sediments. Increased pressure would then affect the

Figure 2.1 The geological time scale

1 Pleistocene
2 Pliocene
3 Miocene
4 Oligocene
5 Eocene
6 Paleocene

TIME	ERA	PERIOD	EPOCH
	Cenozoic	Quaternary	1 2 3 4 5 6
70		Tertiary	
135	Mesozoic	Cretaceous	
190		Jurassic	
225		Triassic	
280	Palaeozoic	Permian	
345		Carboniferous	
395		Devonian	
430		Silurian	
500		Ordovician	
570		Cambrian	
	Precambrian	Ediacaran	

chemistry of the bones which would eventually be turned into stone, but their structure would remain. As the most likely places for the deposition of sediments were lakes and seas, there are more fossils of aquatic organisms than of those which lived on land. The fossils in the sediments below the water may become accessible if there are earth movements which raise the sediments above water level. The land may then be eroded by physical forces or excavated by man.

Sometimes the rock itself is formed primarily from fossils – some types of limestone are mainly made up of the limey skeletons of invertebrates. Although it has been estimated that about 30 per cent of all marine invertebrates of moderate body size *can* leave fossils, the actual fossil record is not this complete. Some layers of sediment may be lost by erosion before the fossils are found. Other sedimentary rocks are so altered by heat and pressure that any fossils they originally contained become unrecognizable. Nevertheless investigations into the strata of sedimentary rocks and the fossils they contained allowed a relative time scale to be developed. The three great eras – the Palaeozoic, Mesozoic and Cenozoic – were named, and then divided into formal 'periods', some of which were further divided into 'epochs'. See figure 2.1.

Dating methods

This relative time scale was mainly established during the Nineteenth Century. At the beginning of the Twentieth Century the discovery of radio-activity gave a method for converting the relative time scale into a reasonably accurate absolute time scale. Each radio-active element decays at a particular average rate. The original **radio-active isotope** (the parent isotope) changes, as a result of radio-active decay, into an isotope of another element (the daughter isotope). If the rate of decay is known, and the relative amounts of the parent and daughter isotopes can be measured, then a calculation of the time when the rock containing the parent isotope was formed can be made. Dating methods involving the uranium/lead decay series were developed in the early years of this century, and other methods such as radio-carbon or potassium/argon dating are now much used. Radio-active carbon ^{14}C is produced from normal nitrogen, ^{14}N, in the upper atmosphere by means of cosmic radiation. The ^{14}C then becomes incorporated into organic matter in exactly the same way as the normal isotope ^{12}C. If the intensity of cosmic radiation can be calculated or estimated, then an estimation of the rate of ^{14}C production may be made. The ratio of

$^{14}C:^{12}C$ in the atmosphere at that time would be similar to that in living organisms because of their constant turn-over of carbon. However, this carbon turn-over would stop as soon as the organisms die, and afterwards there would be a steady reduction in the ratio of $^{14}C:^{12}C$. By knowing the rate of decay of the ^{14}C, the time of death of the organism may be calculated.

Radio-active isotopes may not always be useful for dating particular rock strata as most of the minerals found in sedimentary rocks either are not radio-active or have been derived from much older rocks which were eroded in some way to form the original sediment layers. However, sedimentary rocks often contain segments of volcanic or crystalline rocks which have poured, in a molten state, into cracks and then hardened and which can be dated from radio-active decay.

By means of combinations of these techniques carried out on rocks in many parts of the world, a well defined date has been established for the start of the Cambrian period at 570 million years ago, and the age of the earth itself has been established as 4600 million years. So seven-eighths of the earth's entire history is encompassed in the Precambrian era.

Examples of fossils

The eleven periods of geological time since the start of the Cambrian period are collectively known as the Phanerozoic era – a term derived from the Greek word for visible or evident. In Cambrian strata there are abundant fossils of marine plants such as seaweeds, and of many different phyla of marine invertebrate animals including worms, sponges, molluscs and, particularly characteristic, the early arthropods called trilobites. It was traditionally recognized that there was a dramatic boundary in the record between the Cambrian period with its abundance of fossils and the Precambrian era in which, until recently, it was thought that there were no fossils at all. But life could not have suddenly begun, at a relatively late period in the earth's history, with organisms as complicated as trilobites.

Microfossils
In the early years of the Twentieth Century Charles Walcott discovered a peculiar limestone formation in Precambrian strata in Grand Canyon rocks. The limestone appeared to consist of very thin layers piled one on top of another to make mounds or pillars. Walcott suggested that the mounds, or stromatolites as they were named (from the Greek stroma –

a bedspread, and lithos – a stone), were fossilized reefs which had been formed by various types of algae. Other investigators thought that they were of non-biological origin, and it was not until 1954 that there was further evidence to support Walcott's suggestion. Near Lake Superior, in Ontario, stromatolites were found in an outcrop of Precambrian rocks which contained seams of a very fine grained siliceous material called chert. The chert had been used in the flint-lock guns of the pioneers in the area, and is now known as gun-flint chert. The fossil stromatolites were investigated microscopically after the rocks had been ground down sufficiently for light to pass through them, and the shape of the organisms which had formed them could be seen. Some were almost identical to modern blue-green algae and bacteria. Their shapes had been preserved because a solution of silica had impregnated them after death.

Modern blue-green algae (which are not really algae but are photo-synthetic **prokaryotes**) are found in ponds or in shallow areas of the sea where they form the producer level of many food webs. Living stromatolites have recently been found in a small lagoon, Hamelin Pool, which is part of Shark Bay on the north-western coast of Australia. A sand bar covered with eel grass has reduced the tidal flow at Hamelin Pool and subsequent evaporation has left the water very salty. So the molluscs which would normally be the next step in the food web cannot live there, and the blue-green algae flourish. The cushions and pillars they have built up by the deposition of lime are strikingly similar to the gun-flint fossils.

Well preserved microfossils have been identified in 45 stromatolithic deposits over the past fifteen years and so the early history of life is gradually becoming known. The oldest rocks on earth, formed 3800 to 3900 million years ago, are found in Greenland but they contain no fossils. This could be because they were formed before the earliest life. Alternatively, it could be that any fossils they once contained have been obliterated by heat and pressure. The oldest fossil bearing rocks which have yet been found are 3400 million year old cherts in the Fig Tree Series in southern Africa in which fossil prokaryotes were found in 1977. This find has pushed back the date of the beginning of life as it is a considerable step from the first self-replicating structure to the forma-tion of a prokaryote cell.

Further evidence for a very early beginning of life on earth comes from biochemical studies of a group of organisms, the methanogens, which are conventionally classed with bacteria. These organisms, as their name implies, produce methane from carbon dioxide and they cannot survive in the presence of oxygen. A comparison of some **RNA sequences** of methanogens with those of other prokaryotes of widely different forms (a blue-green alga, a human gut bacterium and a free living bacterium) showed that there was a greater similarity between all the methanogens than between any one and one of the other prokaryotes. It also showed that all the other prokaryotes, although diverse in form, were biochemically more similar to each other than to any methanogen. This suggests that the methanogens form a coherent but separate group within the prokaryotes and, as the early atmosphere was without free oxygen, they may be more representative of the ancestral prokaryote cell. The dating of the blue-green algae and bacteria which form the Fig Tree fossils at 3400 million years old must lead to the conclusion that the common ancestor of the methanogens and the conventional prokaryotes must be much older than this. It is beginning to seem that life appeared rather rapidly (that is, in geological terms where a few million years is rapid) after the production of a solid crust on earth.

The Cambrian 'boom'
The prokaryotes which had already diversified by 3400 million years ago are the only life forms we know about for the next 2000 million years. Towards the end of the Precambrian era fossils of simple multicellular organisms had appeared, such as the jellyfish, soft corals and simple worms of the Ediacara fauna, named after the region where they were found in Australia. By the end of the Precambrian all the major groups of skeletonized invertebrates had appeared, and the stage was set for a great diversification of phyla. So the 'sudden' boom in fossils in Cambrian rocks could be the result of a real increase in diversity. It could also, of course, merely signify that more fossils had been formed because of the evolution of 'hard parts' which were more likely to remain undecayed for long enough to be fossilized.

If it is accepted that a real increase in diversity did occur over about 100 million years in the late Precambrian and early Cambrian, then obviously something must have caused it. **Two explanations** have been suggested. **One** argues that there was a major change in the environment during that time, such as a great increase in the oxygen concentration in the atmosphere. This could have allowed much more complex organisms to survive. However, prokaryotes had been photosynthesizing for a vast period of time before the Cambrian 'explosion' of species, and it seems unlikely that nearly 3000 million years would be required for the necessary oxygen build-up. **The other argument** suggests that a biological rather than a physical limitation had restricted diversification up to that time. The evolution of sex, with subsequent increased variation, is thought to have occurred in the late Precambrian. It is possible that this, allied with the fact of a relatively empty world in which there were many unoccupied **niches**, enabled the Cambrian fauna to branch out in new evolutionary directions. Such a branching out into new directions is called **adaptive radiation** and this diversification near the beginning of the Cambrian is one of the most spectacular examples of adaptive radiation of multicellular life that has yet been discovered.

The relatively sudden appearance of fossils of so many new forms in the Cambrian rocks was known to Darwin when he was writing *The Origin of Species* and he found it difficult to explain. His theory of evolution by natural selection was based on the idea of gradual change, in which species are altered through a very long period of time. This gradualistic view was difficult to equate with the sudden appearance in the fossil record of many different forms of animal life. He suggested that there had been a long, hidden period during which the animals had evolved – a period which was hidden because the sediments formed during the interval were submerged below modern oceans or buried in other inaccessible places. However, he was not convinced that these suggestions could account for the Cambrian 'boom' in fossils, and was generally gloomy about the value of the fossil record in supporting his evolutionary ideas.

It must be remembered that comparatively little was known at the time about the fossil record, and it was the later editions of *The Origin of Species* which contained the more supportive fossil data which Darwin had added. Many palaeontologists in the latter part of the Nineteenth Century found little evidence for gradual change. They were primarily interested in collecting, describing and naming the fossils without necessarily placing them in a Darwinian evolutionary pattern. However, general pathways of evolutionary descent were laid down by the study of some relatively young fossils, particularly the mammal remains from the Cenozoic era. There is a greater chance of finding fossils in relatively young rocks. They are usually nearer the surface than the older rocks, and so are more likely to suffer the erosion which would expose their fossils. Many of the younger deposits are still reasonably soft and extracting their fossils is much easier than trying to break them from very hard rock. The young rocks are less likely to have been deformed by great pressure and heat, which could obliterate the fossils within them. Many of these Cenozoic mammal fossils could be placed in an evolutionary sequence, even though there were still gaps 'waiting to be filled'.

Nevertheless the fossil record does not, in general, produce the long sequence of slightly changed organisms in one particular geographic region which could greatly support Darwin's idea of very gradual change, with a species altering over a long period of time until it must be considered to be a different species. What is more commonly found is that many new species appear at about the same time during a period of adaptive radiation, but each species subsequently seems to change very little. So an evolutionary tree does not necessarily show a direct line connecting present day species with their remote ancestors, but tries to show the pattern of branching which may relate surviving species with those which have become extinct.

Mammal fossils

As we have said, some of the Cenozoic mammals are very well represented in the fossil record. There are numerous fossils of horses which appear to show trends towards increased size, to a reduction in the number of toes, and in an alteration of the teeth to suit a grazing rather than a browsing method of feeding. For a long time the record of these ancient horses was taken as a classic example of gradual transformation. However, further investigation revealed that many of the fossils showed great evolutionary stability. The very early small horses which lived, browsing on leaves, about 40 million years ago lasted through 3 or 4 million years without appreciably changing in form. Similar evolutionary stability is shown by the genus *Equus* in which species have persisted almost without change through the last 2 million years. It seems that the evolutionary tree of horses is an extremely branched bush, in which only one of the branches – the genus *Equus* – has survived to the present time. Many of the other branches survived until relatively recently and showed trends in other directions.

Once a branch line is established, it appears to last for a certain time without much evolutionary change, until it either dies out or gives rise to other branches. It must be remembered, however, that our knowledge of the evolution of horses is based on a limited sample of individuals (which might have been very variable) of each type and which have been fossilized at random. Migrations are assumed to have happened at

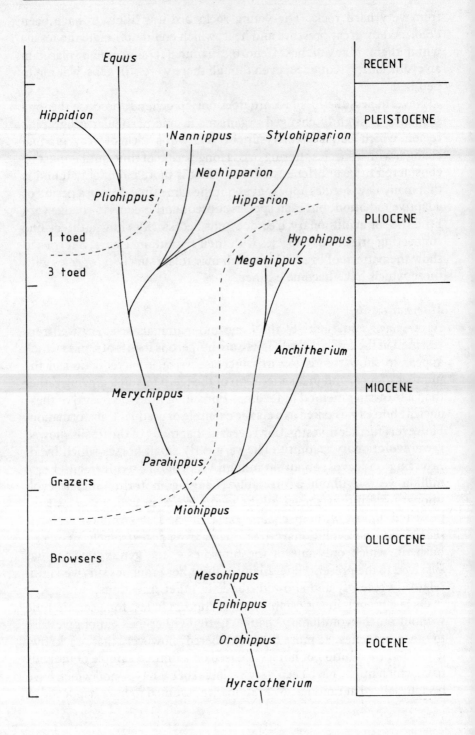

various times to link fossils found in different continents. There is no complete series of fossils in one area through 50 million years to show the evolution of *Equus* from *Hyracotherium*, and decisions about the arrangement of the fossils in the evolutionary tree may be altered as more evidence becomes available.

Gradualism and punctualism

A much branched evolutionary bush such as that suggested for horses has also been proposed to account for the incomplete series of fossils of other groups of related species. Many of the species were changeless over millions of years, and then were rapidly replaced by new species, just as in the history of the horse. This concept of static phases interspersed with sudden changes has been named **punctuated equilibria** or **punctualism**. The idea of episodic rather than **gradual** evolution was discussed by some biologists in the early 1940s. At that time **Ernst Mayr** suggested that some present day genera of birds had evolved rapidly, each as a small marginal population of the ancestral species. In such small and localized populations evolutionary change could be rapid and so, if a similar process had occurred when a fossil series showed that branching had happened, it would be very unlikely that fossils of the actual diversification would ever be found. Mayr's views were very strongly supported in the 1970s by **Niles Eldredge** and **Stephen Jay Gould**. Eldredge and Gould first gave the terms 'punctualism' and 'gradualism' to these alternative views of how evolution might have taken place. As often happens when accepted theories are challenged by new views, a very great polarization developed. Some evolutionists became strong punctualists while others strengthened their gradualist ideas. The debate which went on through the 1970s and which continues at present has led to an assumption by many that the two sides are arguing about whether evolution does or does not take place. What is not always appreciated is that both sides accept that

Figure 2.2 Some of the branches in the evolutionary bush of horses

There is no single upward pattern showing trends towards larger size, more complex teeth and single-toed feet. *Megahippus* was very large, but *Nannippus* very small. *Equus* is the only surviving branch.

species have evolved from other species, and that the debate is about the possible mechanisms which could have brought this about.

The sudden appearance of new species in the fossil record can, of course, be taken by some creationists (who do not restrict themselves to the biblical account) to be evidence of periodic acts of special creation by a supernatural being, but such a view is untestable.

We return to consider the ideas of punctualism and gradualism more fully in a later chapter, in which we discuss how new species may evolve.

BIOGEOGRAPHY

Biogeography is the study of the distribution of organisms. If organisms are adapted to the environment in which they live, one might expect that similar environments in various parts of the world would support similar organisms. But this is not always found. The organisms may show the same adaptations to the environment, particularly if it is harsh, and fill the same ecological niche but they can be of different species, genera or even class. Again one could assume, in an argument unanswerable by scientific means, that all the different organisms had been specially created to fill similar roles in the places where they are now found. One cannot judge why a supernatural being should create marsupial kangaroos as the typical grazing herbivores in Australia and the different types of placental antelopes to fill the same ecological role in Africa.

An alternative explanation, based on evolutionary ideas, suggests that the present day distribution of organisms depends on the movements of their ancestors. The ancestral organisms might have actively moved from place to place, or they could have been passively carried by wind or water currents. A distribution of organisms over a very long time scale could have resulted from the movement of the land on which they lived. If any type of movement led to a certain part of the original population becoming isolated, then each section could continue to evolve along different paths and adaptive radiation could occur. Let us now consider some examples of geographic variation.

The Galapagos Islands

The Galapagos Archipelago is a group of volcanic islands in the Pacific 965 kilometres west of Ecuador. Their name comes from the Spanish

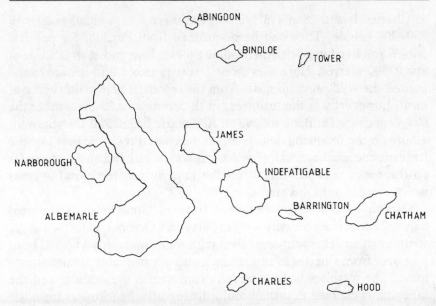

Figure 2.3   The Galapagos Islands
(The very small islands have been omitted from the map.)

name for a tortoise (galapago) but they were also named Las Islas
Encantadas – the Enchanted Isles – by Spaniards who visited them in
1535. The enchantment was not that of their beauty, but because of the
capricious currents around their shores which seemed to attract or repel
ships.

The individual islands bear English names probably given to them by
English buccaneers who made them a headquarters in the Seventeenth
Century. (The Spanish names are given here in brackets after the
English names.) There are five islands which are considerably larger
than the rest and Albemarle (Isabela), the largest island, is about half
the total area of all the islands. Darwin landed on Chatham (San
Cristobal) on 17 September 1835 during his voyage on the *Beagle*, and
the animals and plants which he saw on the various islands were to
prove very important in directing his ideas towards evolution by natural
selection. The *Beagle* sailed round Chatham during the next few days,
and anchored in several bays. The inhospitable nature of the island, the
black lava, leafless shrubs and large cacti impressed Darwin but what
particularly excited him was his first sight of giant tortoises. He also
noted a few dull-coloured birds. On 23 September the *Beagle* sailed on

to Charles Island (Santa Maria) where there was a small colony of 200–300 people. They had been banished from Ecuador for political crimes and lived in a settlement some miles inland and at an altitude of about 300 metres. Here they grew sweet potatoes and bananas, and hunted the wild pigs and goats from the woods. Their main source of meat, however, was the tortoises. In the evening of 29 September the *Beagle* anchored at Bank's Cove on Albemarle Island and Darwin went ashore on the following day. Here on the coastal rocks he first saw the large marine iguana, *Amblyrhynchus cristatus*, which fed on the seaweeds on the rocks. Further inland on the hills he saw terrestrial iguanas which were also herbivorous.

Darwin spent a longer period of time on James Island (Santiago) having been put ashore with a small party on 8 October while the *Beagle* went for water. He spent some time with some Spaniards who had been sent over from Charles Island to catch and dry fish, and to salt tortoise meat. The tortoises were extremely common in the islands, and the inhabitants claimed that they could distinguish the tortoises from the different islands. At first Darwin paid little attention to this claim, and he had already partly mixed together his collection from two of the islands. Later he began to understand the significance of such differences. In *The Voyage of the Beagle*, in which Darwin describes his five year journey around the world, he writes that the tortoises from Charles Island and Hood Island (Española) have shells which are thickened in front and 'turned up like a Spanish saddle', whilst those from James Island have rounder and blacker shells.

Differences were also apparent in the dull-coloured birds which Darwin had already noted on the islands, and which are now called Darwin's finches. The finches seemed related to each other in their colouring, general shape and length of tail, but differed strikingly in the size of their beaks and the type of food they ate.

Darwin wrote 'seeing this gradation and diversity of structure in one small, intimately related group of birds, one might really fancy that from an original paucity of birds in this archipelago, one species has been taken and modified for different ends'. This appeared in *The Voyage of the Beagle* fully 20 years before the publication of *The Origin of Species*, and was the first public indication of Darwin's views on evolution.

Much of our present day knowledge of these small birds is derived from a study made by Lack in 1938. He found that, as in Darwin's time, there were three main types of habitat on the islands:

Figure 2.4   Some Galapagos finches
A   *Geospiza magnirostris*, a large ground finch
B   *Geospiza fortis*, a medium ground finch
C   *Camarrhynchus parvulus*, a tree finch
D   *Certhidia olivacea*, a warbler finch
Based on Darwin.

1   an arid coastal plain with thorn bushes, cactus and prickly pear, with some open ground free from vegetation where lava had recently flowed;
2   higher ground with a higher rainfall which allowed the development of a humid forest. This is found only in the larger islands;
3   open ground above the forest – a habitat which is not colonized by finches.

There are thirteen species of finches on the islands. Because of similarities of morphology and behaviour they are grouped together into a subfamily Geospizinae. There are three genera – *Geospiza*, *Camarrhynchus* and *Certhidia*. Some are ground finches and some live in the trees in the humid forests.

The Galapagos Islands are volcanic, and first began to emerge from below the water of the Pacific Ocean about 10 million years ago. Being volcanic meant that they were, at first, barren of life. Gradually a few organisms landed on them after being carried in water or air currents from the neighbouring continent. The environment they met must

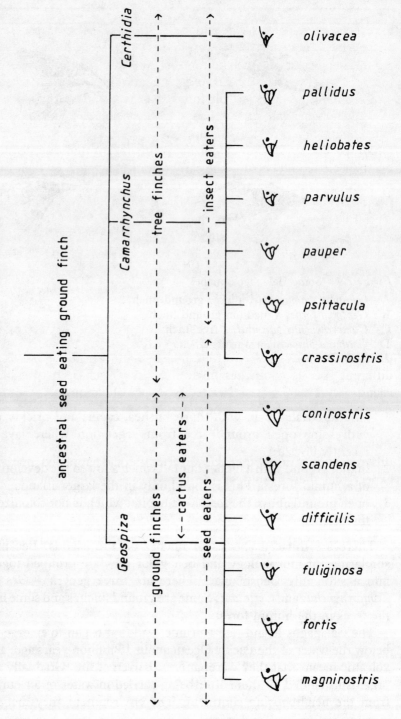

Figure 2.5 Adaptive radiation in Galapagos finches
Seed eating ground finches are thought to have reached the islands and then diversified.

have been extremely harsh, and it is likely that most of the passive arrivals could not survive. However, eventually some plants did gain a toe-hold and managed to take root on the shore, and others survived further up on the slopes of the volcanic cones. Once plants had become established, it became possible for animals to find food and any which arrived, by whatever means, had a greater chance of remaining alive long enough to reproduce. So populations of different organisms could build up. The survival of colonizing plants meant that a succession of plant species could occur – the early colonizers could, by their death, increase the amount of humus in the originally infertile lava. Later plant arrivals would then have a much greater chance of becoming established, and the fertility of the soil would continue to improve. The increased plant cover of the islands meant that there were more ecological niches available for animals. The marine iguanas lived on seaweed around the coast, whereas the terrestrial iguanas which Darwin saw further inland fed on the tough cactus plants. The finches also differed in their habitats and the type of food they ate. The larger islands in the central group could provide a wider range of habitats, having the higher ground which could support the humid forest as well as the more arid coastal plain with its very different vegetation. In these larger islands different species of finches fill the different niches. In the smaller islands, with their more restricted types of vegetation, birds from one species may fill more than one niche.

The variation of species found in the different islands could result from the chance arrival of a particular species on one island, and a different species on another. However, if this had happened, one would expect that there would be identical species of birds on the South American mainland, but the mainland finches all have short straight

Ecological role	Large islands of central group	Smaller islands	
		Charles	Hood
large ground finch	G. magnirostris	G. fortis	G. conirostris
medium ground finch	G. fortis	G. fortis	
cactus ground finch	G. scandens		G. conirostris
small ground finch	G. fuliginosa	G. fuliginosa	G. fuliginosa
humid ground finch	G. difficilis		

Figure 2.6   Distribution and ecological role of some Galapagos finches

beaks for crushing seeds. Some of the Galapagos finches have typically finch-like beaks, although the beaks vary in size enabling the birds to take different seed sizes. Other finches have beaks which are more suited to taking nectar out of cactus flowers, or feeding on buds or soft fruits. Some of the birds are insectivorous, with beaks which enable them to feed on beetles. One of the species uses its beak in a very peculiar way. It fills the ecological niche which is occupied in British woods by the woodpecker, and feeds on insects which live under the bark of trees. The woodpecker manages to ferret out the insects by using its long tongue; the Galapagos woodpecker finch, lacking a long tongue, uses a cactus spine held in its beak for the same purpose.

Continued movement of the mainland finches to the different islands cannot account for the diversity of finches which are found there. It is thought that a few birds, having managed to reach the islands, survived and reproduced. Their offspring then spread out through the whole island group. Even now there are very few species of small passerine (perching) birds on the islands and so there was probably little competition for the ancestral finches when they arrived. Although the original finches might all have had short beaks, it is likely that there would be some range of beak size around a mean. Birds which had identical beak sizes would be competing with each other for the same size seeds. This means that there would be more competition between birds with average beak sizes, and those either above or below the average might be able to get more food. Gradually these birds with the more extreme beak sizes could become more common, and the original finches could have diverged into what became a new species. This process could continue even after the first new species had formed, and the small amount of isolation on the individual islands could maintain the species separation.

On Abingdon and Bindloe Islands, although three species of *Geospiza* are present and each species has a range of beak sizes, there is no overlap. If the large ground finch *Geospiza magnirostris* is not present, as on Charles and Chatham Islands, the food size range taken by the medium ground finch *Geospiza fortis* extends further upwards. On the small island of Daphne and on the Crossman islets only one species of finch is present. Here, in the absence of competition, the range of beak sizes in *Geospiza fortis* and the small ground finch *Geospiza fuliginosa* are very similar.

So gradually beak sizes at one or other extreme of the range for each species may be selected in a process of **character displacement** with a consequent elimination or reduction of competition.

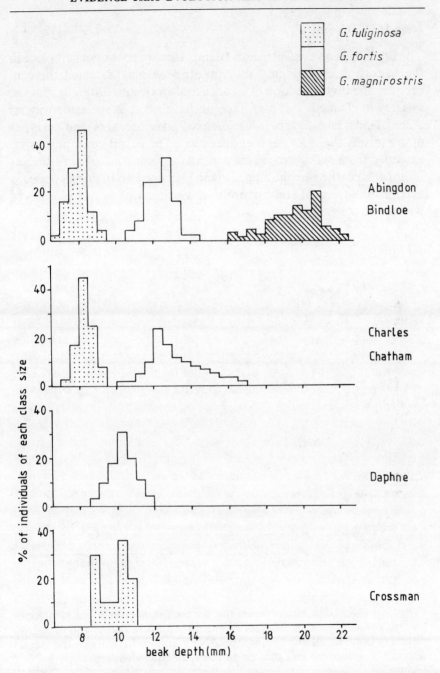

Figure 2.7 Frequency distribution of beak depths in the genus *Geospiza* on the Galapagos Islands, showing character displacement. *Redrawn from Lack*.

Islands in general

The organisms on the Galapagos Islands have a very important place in the history of evolutionary thought. If selection has acted there to produce the diversification of an ancestral stock into different species, as Darwin claimed, it is also likely to have acted in the same way on other islands. Islands vary in size, height and remoteness, and the types of organisms they can support also vary. The island concept may be extended from real islands in the ocean to cover inland 'islands' such as hilltops. Here the surrounding lowland may present, to many species, a barrier to movement and distribution which is just as real as miles of water would be.

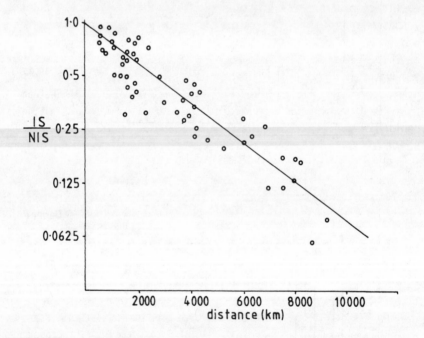

Figure 2.8 Relationship between the number of species of resident non-marine lowland birds on different islands in the South Pacific and the island distances from the colonization source of New Guinea. In each case the species number is compared with that on an island of equivalent area close to New Guinea.

IS – island species number;   NIS – species number on island of equivalent area close to New Guinea. *Redrawn from Diamond.*

Considering islands in the literal sense, however, there are obvious differences between them and their nearest mainland. The water which surrounds the island, apart from being an isolating region, has a stabilizing effect on the island's climate, the scale of which is determined by the island's size. As on the Galapagos, the size of the island may also determine the range of habitats available for colonization by immigrant species. The number of immigrant species depends partly on the distance of the island from the mainland, and also on the prevailing wind and water currents.

Of course the number of species present is also affected by the way in which the island was originally formed. If it is volcanic, like the Galapagos and the much more recently formed island of Surtsey, then it will start off completely devoid of life. Surtsey appeared, and began to build up above sea level on Thursday, 14 November 1963 at a point about thirty kilometres south of Iceland. It is the youngest of a chain of volcanic islands called the Vestmann Islands. Its growth was spectacular. By Friday, the 15th, it was eight metres above sea level; by daybreak on Saturday it had reached thirty-six metres high, and by Sunday it was forty metres at its highest point. One week later the island had a peak ninety metres high. The colonization of Surtsey in the years from its formation has provided a large amount of evidence of what could have happened on other volcanic islands in the past. By 1970 nearly 160 species of insects and some land plants had landed on Surtsey as potential colonists from nearby islands or from the mainland.

In many cases, however, islands have been formed from an existing land mass by the invasion of the sea across a strip of land which became submerged. Some off-shore islands result from the erosion of a belt of soft rock which originally joined them to the mainland. Islands formed by these methods would at first be populated with species which were similar to those on the mainland (making allowance for any difference in habitats which were there before the islands became separated). Some of the species cut off in this way would find no difficulty in moving back and forth over the intervening water. Others, such as animals which could neither fly nor swim or plants which depended for pollination on insect species which do not fly over water, would be restricted for reproduction to those other individuals of their species which had been isolated on the island with them. Because of the small size of some of these isolated populations and the **inbreeding** which would result, certain genetic variations which might have been rare could rapidly spread through the population.

Such small populations which may be founded by a few individuals arriving by chance, or being isolated on a newly made island may be wiped out within a short time, or they may manage to survive and eventually flourish. Some human populations in similarly isolated regions have been closely studied to investigate the effect of inbreeding on the frequency of particular **alleles**. The inhabitants of Tristan da Cunha were brought to Britain after the volcano on the island erupted in 1951. The population was started when a British garrison was posted there during the time Napoleon was held in St Helena (which is 2414 kilometres to the north of the Tristan group of islands). It was thought that there could be an attempt to rescue Napoleon, but in 1817 after his death the British troops were withdrawn.

One soldier, Corporal William Glass was allowed to stay on in the island with his wife and baby son, and they were joined within the next ten years by two more Englishmen, and by three women from St Helena. The three women were a young woman with her mother and aunt. More men (two Americans and a Dutchman) had arrived by 1849, and another St Helena woman settled there in 1863. In 1892 two shipwrecked Italians landed, and in 1908 two Irish women emigrated there. By 1961 there were 270 Tristanians who were descended from these pioneers. This very inbred community showed a much higher incidence than that normally found of the inherited disease, *retinitis pigmentosa*, in which a chronic inflammation of the retina leads to blindness. Four of the 270 people had the disease, which is caused by a **recessive gene**, and it is probable that they all had common ancestors. So an allele may spread through a small population relatively rapidly because of inbreeding.

The same type of effect is seen when an established, isolated population is drastically reduced by some natural catastrophe. Pingelap is a small cluster of islands in the Pacific. In the late Eighteenth Century all but about thirty Pingelapese were killed by either typhoon or by the subsequent famine, and the present population of 1600 people are descended from these thirty survivors. About 5 per cent are **homozygous** for a rare form of partial blindness called *achromatopsia* although it is likely that only one of the original thirty was **heterozygous** for this rare, recessive allele. So there has been a rapid increase in the frequency of the allele.

In such small, isolated communities which may be formed by limited immigration or by natural catastrophe, each organism is restricted in its reproductive 'choice'. The characters shown by an organism (its phenotype) depend on the program encoded in its genes (its genotype) and also on the environment. During sexual reproduction there is a great shuffling of alleles during **segregation** and **genetic crossing-over**. The number of new combinations which can be produced is determined by the frequency of various alleles in the parent population. It is rather like setting out for an afternoon's sketching and painting and losing all the tubes of red and yellow paints on the way. The finished painting, from this restricted palette, may have very different colour qualities from those you planned. Of course, you might have produced a masterpiece – the first painting of your new 'blue period' – but it is more likely that the overemphasis of your remaining colours will have been unsatisfactory.

An immigrant population or a population very much reduced by a catastrophe has a restricted range of alleles and the result of shuffling this restricted range is like painting the picture with few colours. If a few of the founding individuals are heterozygous for a deleterious recessive allele, the small population makes it fairly likely that these heterozygotes might mate. This means that the chance of producing affected homozygotes will be relatively high. It is extremely unlikely that the few individuals which might start a new population will carry the complete range of alleles found in the original population. This has been summarized as the **founder effect**, a term introduced by Mayr. So the original population, shuffling its wide range of alleles each time sexual reproduction occurs, and the isolated population, shuffling its restricted range, are likely to diverge from each other during subsequent generations.

Continental distribution

Biogeographic realms

Nineteenth Century naturalists like Alfred Russell Wallace recognized that there are biogeographic realms which correspond only roughly with modern continents. Each realm has a characteristic group of organisms. See figure 2.9.

The native mammals of Australia are non-placental. Many are marsupials, but there are also monotremes such as the duck-billed platypus. These native mammals filled the available ecological niches. In other areas similar niches were filled by later placental (eutherian) mammals. The mammals, whether marsupial or placental, which filled similar niches became strikingly alike. So the marsupial 'cat' *Dasyurus*

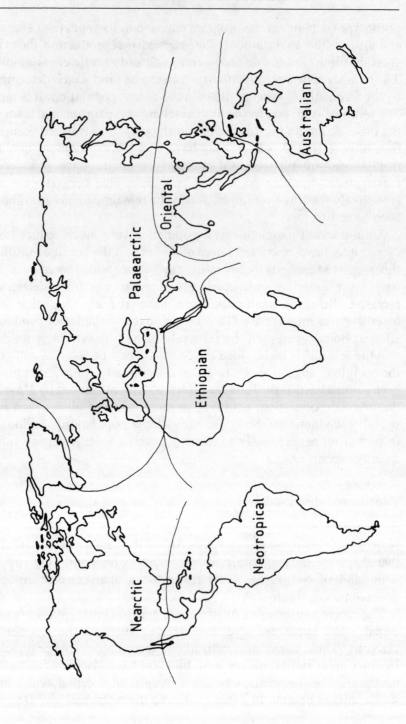

Figure 2.9   Major zoogeographic realms of the earth

looks like the placental cat *Felis*; the marsupial 'mouse' *Dasycercus* is like the placental mouse *Mus*; the marsupial flying phalanger *Petaurus* is similar to the placental flying squirrel *Glaucomys*, and the marsupial 'wolf' *Thylacinus* resembled the placental wolf *Canis lupus*. However, even though the resemblance may be striking, all the marsupials are more closely related to each other than to any placental mammal.

The presence of a pouch is not the result of adaptation to a special set of conditions on the Australian continent, but of a historical event – the isolation of the continent from the main stock of mammals. The marsupials are not the *best* mammals for Australia; placental mammals have been extremely successful there whenever they have been introduced, usually by man. Although Australia is the main region, it is not the only part of the world where marsupials are found. Phalangers and cuscuses

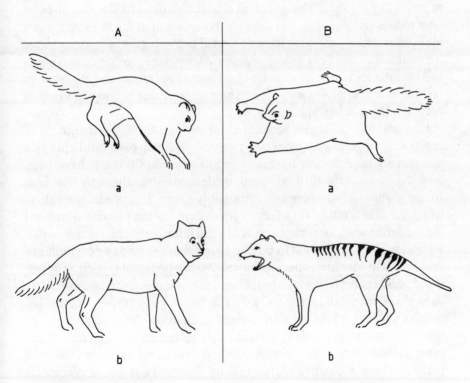

Figure 2.10 Convergent evolution of Australian marsupials and placental mammals of other continents

A   placental mammals   *a* flying squirrel   *b* wolf
B   marsupial mammals   *a* flying phalanger   *b* Tasmanian 'wolf'

are found not only in Australia and Tasmania but also in New Guinea and Celebes. Many species of opossum are found in South and Central America, and one species ranges into the southern parts of the United States.

South America, apart from being a secondary region for marsupials, has a large number of unique placental mammals such as edentates (sloths, anteaters and armadillos), caviamorph rodents (chinchillas, cavies and capybaras) and platyrhine monkeys with prehensile tails. In the past South America was also an isolated continent, the Isthmus of Panama having been formed only a few million years ago.

Certain types of animals, however, are found in different isolated biogeographic realms. Of the three genera of modern lung-fishes, one is found in South America, one in Africa and one in Australia. Similarly the flightless birds, the ratites, are widely distributed over the continents. They all share the common characteristics of a flat breast-bone which does not provide a large enough surface area for the attachment of flight muscles, and also have powerful heavy legs which are suitable for running. The rheas in South America, ostriches in Africa and emus in Australia and New Guinea are similar in appearance, the cassowaries of Australia and New Guinea looking rather different and the kiwis in New Zealand being much smaller.

Classical biogeography accounted for the present day distribution of organisms in terms of dispersal from an original place and this is a satisfactory explanation for many animals. Australia *could* have been invaded from Asia through land bridges passing through the East Indies – the distances are not impossibly great. But one of the things which struck Wallace very forcefully during his long journeys around these islands was the abrupt change in the fauna between islands which were very close together. On Bali he found that the birds were similar to those he had already collected on the Malay peninsula, such as green woodpeckers and barbets, whereas on Lombok, separated from Bali by only thirty-two kilometres of sea, the birds were typical Australian species such as white sulphur-crested cockatoos. As he continued to travel around the islands (see figure 1.3) he found that he could place some island faunas into the Australian and some into the Oriental realms. There are native marsupials on Celebes, but native placentals on Borneo. The division snakes its way between the islands in what is now known as the Wallace Line. If all the Australian fauna had moved there across South-east Asia, one might expect to find that all the islands had similar animals.

Continental drift and plate tectonics

The remarkable similarity of the facing coastlines of South America and Africa led to a suggestion in the Seventeenth Century that the two continents had, at one time, been closely linked together. In 1911 Alfred **Wegener** produced a **theory of continental drift** based not just on shape but also on close geological similarities between the two continents. He proposed that until about 200 million years ago the land on earth was in a single huge continent which he called Pangaea. Pangaea subsequently broke up and the pieces drifted apart to form the present continents. However, the forces which could have caused such drifting were unknown and the theory, although attractive in many ways, lost support. In the 1960s the discovery of particular patterns of magnetism on the ocean floor led to the theory being re-examined. It is known that there is a world wide series of cracks in the floor of the oceans, and that molten material continually exudes through these cracks. The molten material spreads out from the crack, and very rapidly cools and hardens. See figure 2.11.

So new ocean crust (the lithosphere) is constantly being made on each side of the crack, and is moving apart at a rate of a few centimetres a year. The position of the cracks divides the surface of the earth into various **plates**. Continents, which are made of lighter material than the ocean crust, are therefore carried along on the moving lithosphere beneath them. So the American continent is gradually moving away from Europe and Africa as the Atlantic Ocean slowly widens. In various regions of the earth, the reverse process occurs. There, where two plates which are moving in opposite directions meet, one of the plates sinks down underneath the other in a process of subduction. The material in the sinking plate remelts to form the hot magma from which new lithosphere may eventually be reformed. The regions in which subduction occurs are likely to have volcanic activity, and many of them are also earthquake zones. Volcanic activity is also likely to occur along the crack lines. Of course, if continents meet each other as a result of the moving together of the edges of the plates which carry them, subduction cannot occur. Some of the continental material is then pushed upwards to form great mountain ranges.

Wegener's theory can now be reconsidered in the light of **plate tectonics**. The supercontinent Pangaea is thought to have existed about 225 million years ago in the Permian period. It started to split up in the Triassic. Pangaea was a land mass which stretched almost from the North to the South pole, surrounded by the universal ocean

1

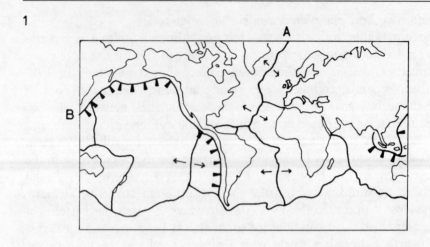

2

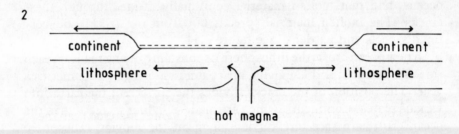

3

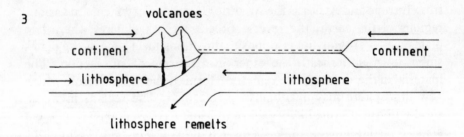

Figure 2.11

1   The major plates of the earth's surface and the direction in which they are moving. The map shows the positions of the cracks in the bed of the ocean where new lithosphere is being formed by an upwelling of hot magma from the earth's interior. The new material spreads out on each side of the crack. In other regions the lithosphere from one plate sinks underneath that of another plate (subduction). The direction of this sinking is shown by the arrow heads.

2   Spreading zone – as at A

3   Subduction zone – as at B (for fuller description, see text)

Figure 2.12   The possible shape of Pangaea 200 million years ago, showing the positions of the potential modern continents

Panthalassa. The Tethys Ocean was an inlet in the irregular outline of the supercontinent.

The first main split opened up the North Atlantic Ocean and the second formed the Indian Ocean. The northern land masses, Laurasia, then moved away from the southern masses, Gondwanaland. So South America, Africa, Antarctica, India and Australia were, at one time, part of the same land mass. India then moved upward to collide with Asia and to raise the Himalayas. Africa, which separated from South America, rotated and closed the Tethys Ocean which remains as the Mediterranean. Australia rifted from Antarctica and moved to its present position.

The geographic distribution of organisms is now, therefore, more explicable. The presence of ratite birds, or lungfishes, in South America, Africa and Australia could date from the time when the areas were joined together. Continued evolution could produce the differences in present day forms. The change in fauna along the Wallace Line depends on the different origins of the Oriental and the Australian realms.

Clinal distribution

So far we have mainly considered species which are now widely separated. However, there is a different form of geographic variation. Species may vary slightly across their geographic ranges, becoming adapted to the local conditions. Such a more or less continuous change in the characters of a population is called a **cline**. Some clinal variations seem to be obvious adaptations to the environment. Endothermic animals which rely on maintaining a constant body temperature by means of metabolically produced heat (warm-blooded animals) may have problems of heat conservation in a cold environment. They tend to be larger in these regions, thereby having a smaller surface area/volume ratio. This is summarized as **Bergmann's Rule**.

They also have smaller and shorter projecting parts such as tails, beaks and ears (**Allen's Rule**). There are also clinal colour variations; animals tend to be darker with more melanin pigmentation in warmer regions (**Gloger's Rule**).

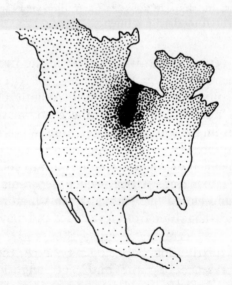

Figure 2.13 Clinal variation in the size of house sparrows in North America – the darker the shading, the larger the birds. The birds are larger in the coldest regions. This illustrates Bergmann's Law. *Based on Johnston & Selander.*

The amount of change over short distances is small but the forms from the extreme edges of the geographical range may be so different that they could be classed as subspecies or even as different species. Usually the extremes do not meet, but if the geographical range is very extensive the extremes may inhabit the same area, thus forming a **ring species**. The herring gull, *Larus argentatus*, and the lesser black backed gull, *Larus fuscus*, behave as separate species in Britain and do not interbreed. Fertile hybrids can be produced, however, if eggs are interchanged between nests. The fostered offspring mate with a member of the other species when they mature. It has been shown by Stegman that there is a chain of ten recognizable forms or subspecies of gulls belonging to the *fuscus/argentatus* group which form a ring around the North pole. Westward from Britain across North America the gulls look more like herring gulls. Across the Bering Strait and into Siberia they begin to look more like black backed gulls. The two forms found in Britain are the ends of the ring which happen to overlap. A similar ring is found in the distribution of the great tit, *Parus major*. The British species is part of a series of forms which extend eastward through Europe, Iran, India and South-east Asia and then northward through China. A second series extends from Europe north of the deserts of southern Russia to Mongolia and north China. The terminal links overlap in the Amur Valley of north China where the two types can interbreed.

COMPARATIVE ANATOMY

On the basis of the similarity of characters organisms can be grouped into species; species into genera; genera into orders and so on. This, in itself, is not an argument for evolution as anything can be placed in a hierarchical arrangement. Carl von Linné, the Eighteenth Century Swedish naturalist whose name was latinized to Linnaeus, catalogued a wide range of organisms into genera and species. He believed that a classification should show God's idea of order. The construction of a modern biological classification takes into account the order of descent of organisms from common ancestors, and also the extent of divergence from these common ancestors. As in many other sections of evolutionary thinking, there is a lively debate going on at present about the 'best' method of classifying organisms, and we return to this topic in chapter 6.

Homologous and analagous structures

Characters which have been derived from a common ancestor, **homologous structures**, may be so unalike in their present day appearance and function that their homology may be shown only by dissection. Other characters which may look extremely similar in two organisms, **analogous structures**, may owe their resemblance to convergent evolution rather than to common ancestry. If one accepts that, during the course of evolution, organisms become more adapted to the local conditions in which they live, then one would expect that there would be a divergence from a basic ancestral pattern for organisms living in different environments, and a convergence from several ancestral patterns for organisms living in similar environments. Both the divergence and convergence depend on changes in structure and one assumes that the end product of the structural change gives a selective advantage. But what of the stages in between? In 1871 St George Mivart argued that Darwinism must fail because the theory was unable to explain the incipient stages in the evolution of useful structures. The feathers of birds would enable them to fly and exploit the air. But what is the use of a reptile scale which has changed only 5 per cent towards being a feather? There would be no flying advantage at all, and so how could feathers have arisen by selection? There have been two basic answers to this, and similar questions.

Pre-adaptation

The term pre-adaptation might suggest some form of plan or foreknowledge of what might be useful in future circumstances, but it really means that a structure in which the original function is becoming less important might therefore be available for use in a new form. *Archaeopteryx*, a small animal which was about the size of a crow, is thought to be typical of the ancestors of birds. Its skeleton was very similar to that of bipedal ground living dinosaurs which lived at the same time, but fossils show that it possessed feathers. If *Archaeopteryx* was endothermic, then the change of reptile scales into feathers, even if they were originally very small, could have provided some insulation and given a selective advantage. An increase in the size of certain feathers, while not affecting insulation, might have given the animal other advantages. One theory suggests that the trailing larger feathers which were present on the fore limbs were used to form a net to catch flying insects while *Archaeopteryx* ran along the ground with the limbs extended. If the animal then started to jump in the air to catch more insects, a selective advantage could be seen in the development of true flight feathers.

A change of function rather than a completely new structure is also seen in the **evolution of jaws**. The earliest fish had no jaws, but did have a series of bones supporting the gill arches. The most anterior gill arches, the mandibular arches, were in a suitable position to be transformed into jaws, although they had worked perfectly satisfactorily as gill supports.

During the evolution of jaws the second of the gill arches, the hyoid arch, became the hyomandibula, a bone which connected the new upper jaw with the skull. This bone remained as a jaw support in fish, but in terrestrial vertebrates it evolved into a bone for the transmission of sound. Amphibians, reptiles and birds have only this single ear ossicle, the stapes, but mammals have a further two bones, the malleus and the incus, to transmit sound vibrations across the middle ear. If these two 'new' bones had to evolve gradually from scratch, Mivart's objection would stand. However, evidence from embryology and palaeontology shows that the 'new' bones have a different origin. In

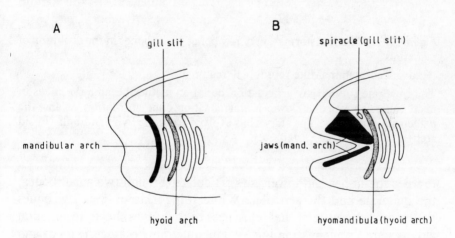

Figure 2.14   The possible evolution of gill arch bones into jaws

A Jawless vertebrate   B Jawed fish

The ancestral jawless vertebrate supported the gill slits by a series of gill arches. The first gill arch, the mandibular arch, became the jaws and the second gill arch, the hyoid arch, became the jaw suspension in the original jawed fish.

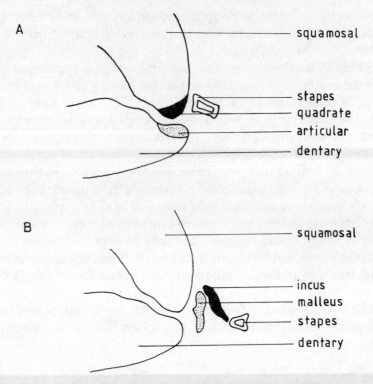

Figure 2.15   The transformation of jaw bones to ear bones in the evolution of mammals

A advanced mammal-like reptile   B mammal

The quadrate (upper jaw articulation bone) of reptiles became the incus of mammals. The articular (lower jaw articulation bone) of reptiles became the malleus of mammals. The stapes has evolved from the hyomandibula (hyoid arch) in all terrestrial vertebrates.

reptiles the upper and lower jaws articulate through two small bones, the quadrate and the articular. While still functioning as jaw bones these could, because of their closeness to the stapes, begin to function also as sound transmission bones. The quadrate became the incus and the articular became the malleus of the mammalian ear.

So there was no stage when the bones which would make the jaws and those which would make the ear ossicles were not functioning either in their original or their evolved pattern. Evolution could just convert one function into another. Tracing the pattern of homologous structures found in different organisms can show how the structures have become altered to fulfil different functions. Consider the common example of homologous structures – the **pentadactyl limb** of vertebrates. The basic pattern of the bones can be traced through the amphibians, reptiles, birds and mammals, although the relative proportions of the bones may differ and some bones may be reduced or absent. The limbs may be specialized for swimming – as in the flippers of the whale or the wings of penguins; for walking – as in the legs of terrestrial vertebrates; for manipulation – as in the fore limbs of primates; for flying – as in the wings of birds and bats. Adaptation for the same mode of life may be achieved by different means; bird wings, for example, depend on feathers whereas bat wings use a membrane of skin. The skeleton beneath, however, is remarkably similar.

The similarity between the skeleton of the wings of birds and bats suggests that they have both evolved from some common ancestral

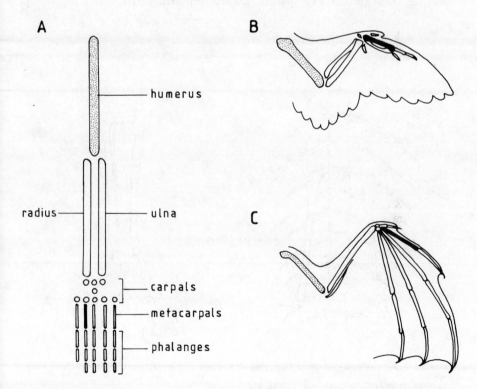

Figure 2.16 Comparison of the skeleton of bird and bat wings with the typical pentadactyl limb

A pentadactyl limb   B bird wing   C bat wing

terrestrial vertebrate. The appearance of flight in mammals was a very much later development than flight in birds. Bats evolved to exploit a way of life unfilled by other mammals. So flight in birds and bats does not demonstrate recent common ancestry between the two – the wings have evolved independently in each. There are probably better designs available to a creator of wings from scratch – think of all the different types of aeroplanes and other flying machines (fixed wing, delta wing, changed geometry wing, helicopters) – but the bird and the bat wing have had to use the available starting plan of the homologous pentadactyl limb.

Homology between structures can also help to explain the presence of **vestigial organs** – structures which were once useful but which are now without function. Snakes have rudimentary leg bones which no longer help them to move, and horses have non-functional toe bones which, having been reduced to splints, do not touch the ground.

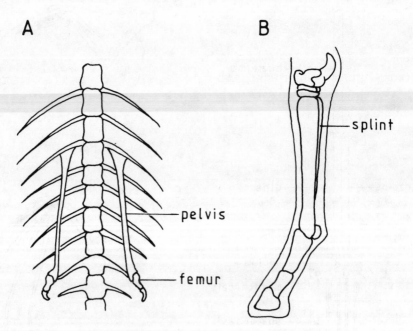

Figure 2.17 Vestigial structures as evidence for evolution
A snake pelvic region   B horse leg
Snakes have evolved from 4-legged ancestors and some, such as boas and pythons, retain very small leg and pelvic bones. Horses retain vestiges of side toes in the splint bones.

Vestigial structures can therefore be seen as those which are on the evolutionary path to further and further reduction and eventual loss because their presence gives no selective advantage to the organism possessing them.

Rapid transitions

It is possible that a full sequence of intermediate stages was not needed in the change of a structure from one form and function to another. Some major changes could have arisen fairly rapidly as a result of altered development rates for different structures during embryological life. The development of an organism depends on the program encoded in its genotype which is 'read' during embryology and later life as the genotype is translated into the phenotype. If the program is read as a series of small steps, a small change in an early step can lead to profound changes later on. Consider the relationship between man, *Homo sapiens*, and his closest living relative – the chimpanzee, *Pan troglodytes*. As you see, these two animals are placed in different genera although it has been found that they are more alike in the structure of their genes (even if their chromosome structure differs) than any pair of species within a single genus which have so far been studied. So what has caused the large morphological differences between the two?

Our direct knowledge of their genotypes comes from comparing the amino acid sequences of their proteins – proteins which have been made because of the coding of **structural genes**. But other genes, **regulator genes**, control the mode of action of these structural genes. **Transcription** of the regulator genes may produce RNA molecules which act as, or act with, effector molecules which can either switch on or switch off certain structural genes. Depending on the rate and pattern of switching, enzymes needed for development may be produced at very different rates in different animals. There is some evidence that this happens in the development of the chimpanzee and man. Embryo chimpanzees and humans look very similar, but then human growth slows considerably. Although the direction of growth change is similar in the two animals, the extent of change is much less in man. This is particularly clear in the growth of the skull. So the large differences between adult man and adult chimpanzee may be the co-ordinated response to a single change in rate. See figure 2.18.

This means that an alteration in development rates could occur between different organ lines within the same animal, leading to the development of some adult characters whilst the animal retained a

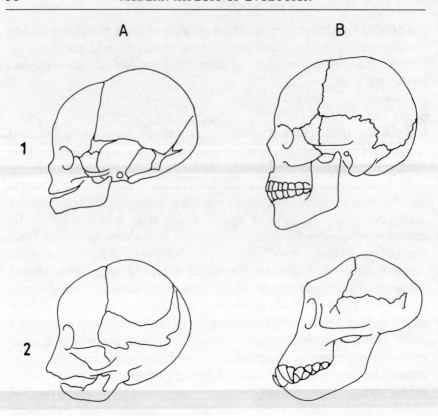

Figure 2.18   Comparison of adult and foetal skulls in man and chimpanzee
A foetal   B adult
1 man   2 chimpanzee
The foetal skulls of man and chimpanzee are very similar. During growth to adult life both skulls change but the chimpanzee skull changes much more from the foetal form.

basically juvenile form. The Aztec Indians collected animals, which they named axolotls, from Lake Xochimilco for food. These animals puzzled Europeans; there was a strong resemblance to salamander tadpoles, but these 'tadpoles' were sexually mature. Then one of the 'tadpoles' transformed as a result of fouling of the water in which it lived – it 'grew up' into a normal salamander which could be placed in the genus *Ambystoma*. So the juvenile stage of ancestors had become the adult stage of the descendents, a process which has been named **neoteny** or **paedomorphosis** (child-shaped). Some of these paedomorphic salamanders have retained the ability to metamorphose into the ancestral adult form if environmental conditions are changed. Others, the perennibranchs (permanently gilled) are permanently juvenile.

Paedomorphosis is therefore a way in which large differences can result from relatively small alterations in timing, and many evolutionists think that it has played an important role in the evolution of several large groups of animals. The phylum Chordata is a large, diverse group of animals comprising four subphyla. Three of these are often referred to collectively as protochordates (lower chordates); the fourth subphylum, the vertebrates (higher chordates) contains all animals with backbones. All chordates at some stage of their life history have pharyngeal pouches which may become gills, a dorsal nerve cord and a supporting skeletal rod named the notochord. The protochordate tunicates (sea squirts) show all of these features when they are larvae and at that stage look very much like amphibian tadpoles. However, at metamorphosis the tail disappears and the animal rapidly changes its shape and becomes sedentary. If sexual maturity could have been reached by a genetic change in development rate while the ancestral tunicates were still in their motile, larval form then evolution into vertebrates could have occurred without a long line of intermediate forms.

Comparison of immature forms of different animals, and of the larval stages of one group with the adult stages of another, has helped to establish possible evolutionary relationships between them. As we have mentioned, all chordates have pharyngeal pouches at some stage in their lives. They become functional gills in those vertebrates which primarily live in water, such as fish, and in the larval stages of amphibians. However, they are not of any value to terrestrial vertebrates which must enclose their delicate respiratory surfaces more efficiently, both to prevent them from drying up and so becoming useless and to protect them from damage. Lungs, into which air could be pumped, were formed from a new outgrowth from the alimentary canal and these took the place of the gill slits through the alimentary canal wall which function well for the one-way movement of water in fish. However, the basic chordate pattern of pharyngeal pouches is still apparent in these terrestrial vertebrates, even if the pouches have different functions. One is involved in the formation of the Eustachian tube connecting the middle ear with the pharynx, and others with the palatine tonsils, the thyroid, the parathyroids and the thymus.

BIOCHEMISTRY AND PHYSIOLOGY

All of these changes in structure and function are, of course, dependent on alterations in the biochemistry of the cells involved. There is a fundamental similarity in the basic biochemical reactions which occur in organisms. All the different forms from the smallest micro-organisms to the largest and most complicated plants and animals use, as a basic currency, a very limited range of molecules – about twenty different **amino acids**; five **organic bases** of two types, **purines** and **pyrimidines**; the **pentose sugars ribose** and **deoxyribose**, and other carbohydrates; **lipids** for membranes; **phosphates**. Of course, such a small basic molecular pool could be seen as evidence of the economy practised by a creator who chose to work with a limited number of materials from which to make the different organisms. It is also possible that only these compounds function efficiently to support life as we know it. To begin with many many different organic compounds might have been formed from simpler materials, but only some of these compounds were to acquire some of the properties of life.

Origin of organic compounds

The way in which such organic compounds might have been originally formed from inorganic material was suggested by the results of the Miller and Urey experiments carried out in the University of Chicago in the 1950s. A sterilized flask containing water vapour and a mixture of the gases methane, ammonia and hydrogen had high frequency spark discharges passed through it for a week, during which time the water was kept boiling in another flask to produce the water vapour. At the end of the week the materials formed in the flask were analysed. Milligram quantities of the two amino acids alanine and glycine were found with smaller quantities of other amino acids, fatty acids and other organic compounds. Since then large numbers of other experiments have been carried out using a variety of different gas mixtures in the flasks to simulate possible atmospheric conditions. The energy sources have also been varied. It has been found that in reducing atmospheres, such as the original mixture used by Miller, moderate energy sources are sufficient to produce most of the classes of organic substances needed for life. In oxidizing atmospheres, much more energy is needed to produce the same results and the products tend to be rapidly oxid-ized. It is difficult to decide which of the many different gas mixtures might represent conditions on the early earth. In the same way, one cannot be sure which energy source was the most important. Ultraviolet radiation from the sun could have penetrated the seas to a depth of some metres, and could have produced cyanide (HCN) if ammonia was dissolved in the water. Although some of the products would be unstable and would tend to break down again into their constituents,

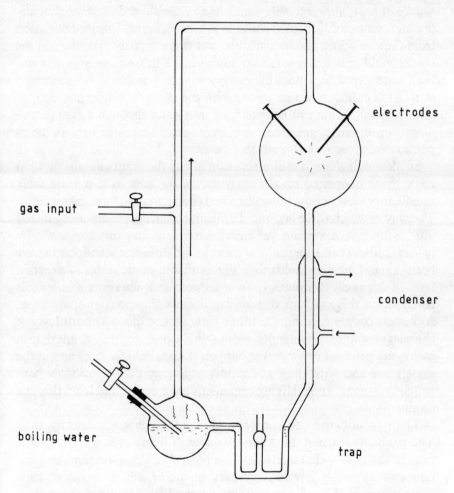

Figure 2.19   Miller's apparatus

A spark discharge is maintained in a flask which contains a mixture of methane, ammonia, hydrogen and water vapour, the water vapour being constantly supplied by boiling water in another flask. Materials can be removed from the trap for analysis.

many could accumulate. Ultraviolet radiation, being part of the sun's radiation, would be continuously available during daylight. Other sources of energy were much more intermittent. Lightning, which was simulated in Miller's experiments, might have been very important at certain stages. Heat energy can be used to synthesize organic compounds from inorganic materials in the laboratory, but its role on the early earth is controversial. Volcanic activity could provide a natural source of heat, although this might be so variable and extreme that any organic compounds formed could break down again. Infrared radiation from the sun would also be available, but as it is rapidly dissipated in the seas it could only act in small shallow pools. Ultrasonic energy has also been considered as a possible energy source – it could be generated during the collision of a meteorite with the earth. In one experiment to mimic this situation a bullet was fired into water through a gas mixture of carbon dioxide, ammonia and hydrogen. Traces of organic compounds were recovered from the water.

So although there is still discussion about the actual details of what could have happened on the early earth, we now have a reasonably satisfactory view of how these chemical changes could have taken place. The way in which these organic chemicals could have become organized into living systems is not yet clear. Assuming that this happened by natural rather than supernatural means, two different standpoints have been taken. One view holds that 'life' is an automatic result of a certain level of chemical complexity. So whenever and wherever this level is reached, on this earth or outside it, 'life' would spontaneously arise. Although people holding the other view accept that a certain level of chemical complexity is needed before 'life' can arise, they think that so many independent events must happen at just the right time and in the same place and with the right supply of energy that it probably happened only once, and all living organisms have descended from this one origin.

There is, of course, the problem of deciding what is meant by 'life'. You probably learned the 'characteristics of living organisms' at some time in the past – characteristics like reproduction, nutrition, respiration and so on. It is relatively easy to define life in terms of such characteristics. What is not easy is to say at which point in the transition from non-living chemical systems to living systems the big change-over occurred. Some investigators consider that any self-replicating system, even if it is only a self-replicating molecule, should be thought of as alive and that nucleic acid synthesis was therefore the first step. Others feel that energy-deriving and energy-using processes came first. Living cells need both of these abilities. They must be able to metabolize – to rearrange the atoms of the materials they ingest into molecules they need to ensure their survival. They must also be able to reproduce and so pass on these biochemical abilities to their offspring. Whether the metabolism capacity came before the reproduction capacity or vice versa, or whether the two evolved simultaneously is open to speculation. It is a little like asking whether the chicken or the egg came first. The genetic code cannot function or replicate without enzymes, but enzymes cannot be made without the genetic code. However, recent research in the USA has shown that some nucleic acids may themselves act as enzymes. One form of RNA can act as an enzyme in trimming transfer RNA molecules and preparing them to transport amino acids to the **ribosomes** for the production of proteins. If RNA molecules can act in this catalytic way now, it is possible that they could also do so in the earliest self-replicating systems. Many investigators are presently involved in examining the possibilities for the evolution of the main characteristics of living systems – being bounded by a membrane, controlled metabolism and replication – from simple organic systems.

Even though we may assume that this could have happened in the past, it does not follow that it is a process which occurs today. Changes in the chemistry of photosynthesis produced a major change in the conditions on earth. The first type of photosynthesis probably used hydrogen sulphide as a supplier of the hydrogen needed to reduce carbon dioxide into carbohydrates, and this reaction still occurs in present day green and purple sulphur bacteria. The use of water as a hydrogen source became a much more widespread pattern of photosynthesis. The sulphur produced by the first method accumulated on the earth; the oxygen produced by the second method accumulated in the atmosphere. Some of it, in the outer atmospheric layers, was converted into ozone and this radically reduced the amount of ultraviolet radiation which could reach the surface of the earth. The non-biological synthesis of organic matter was then much less likely to occur, and the atmosphere was converted into one in which one-fifth of the molecules were oxygen. Organisms could then take an evolutionary path whereby this oxygen could be used in reactions from which they could obtain energy for other metabolic processes.

Comparison of amino acid sequences in different organisms

Even though the basic chemistry of metabolism and reproduction is similar in all organisms, there are differences in the details of the processes. Proteins are made under the coded instructions of genes and so there may be alterations in their amino acid sequences as a result of gene mutations. Mutations are spontaneous events which can happen at any part of the genetic apparatus. The number of changed amino acids in a protein which is comparable in different organisms could therefore give some idea of when the organisms evolved separately from a common ancestor. One advantage of this study is that the information is more readily quantifiable than that derived from comparative anatomy. It is also possible for very different organisms, such as a fungus and a mammal, to be directly compared. Fitch and Margoliash have produced a **phylogenetic tree** for twenty different organisms based on

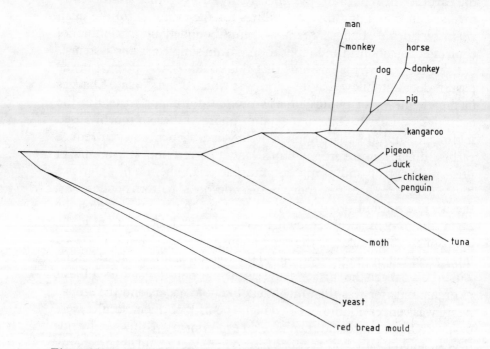

Figure 2.20   Evolutionary tree for cytochrome c

Differences in the amino acid sequences of cytochrome c were used to produce a computer generated tree. This agrees reasonably well with evolutionary relationships suggested by the fossil record. *Adapted from Fitch and Margoliash*.

amino acid substitutions in cytochrome c, a protein involved in cell respiration. The phylogeny agrees reasonably well with the relationships between the organisms which can be inferred from the fossil record.

Comparison of hormones in different organisms

Evolutionary chemical relationships may also be traced by studying hormones in different groups of organisms. Hormones are ubiquitous and, from an evolutionary point of view, extremely ancient compounds. Medawar has pointed out that endocrine evolution is not an evolution of hormones but an evolution of the uses to which they are put. Certain hormone configurations such as the **steroid nucleus** are found in most species of animals and in many species of plants. Most hormones belong to 'family groups', the members of which are related in a way which suggests an evolutionary connection. In some cases the members of a 'family' retain a common biological activity; in others their biological functions have diversified. The pituitary hormone prolactin, for example, reduces the permeability to sodium ions of the gill membrane of bony fish; it stimulates certain newts to return to the water to breed; its secretion results in the production of 'milk' in the crop of pigeons and doves for feeding the young; it promotes milk ejection in lactating mammals. Thyroid hormones are important for differentiation in amphibians, birds and mammals, but also for metamorphosis in amphibians, and for producing heat in mammals. Even those hormones which retain the same function such as the insulins, the gonadotrophic and adrenocorticotrophic hormones may show changes in amino acid sequences in different organisms. The effect of the change, however, is limited by the position of the altered amino acid. Mammalian adrenocorticotrophic hormone, ACTH, is a polypeptide made of thirty-nine amino acids, the sequence of which has been determined in a number of species. It appears that all the amino acid substitutions are in the last part of the chain where they do not seem to affect the corticotrophic effect of the molecule.

The combination of all these biochemical, physiological and structural similarities and differences between organisms can provide very telling evidence for evolution.

OBSERVED EVOLUTION

Some changes in organisms have occurred during a relatively short time span. Some of these changes have been deliberately sought by man for one reason or another; others have occurred naturally.

Artificial selection

Darwin decided on the term 'natural selection' as an analogy to the artificial selection which had been practised for so long by animal and plant breeders. More than 4000 years ago the Egyptians carried out controlled breeding of their domesticated cattle, and had selected certain food plants. Selection for various characteristics in plants and animals has gone on ever since. In England in the Eighteenth Century pioneer animal breeders such as Robert Bakewell established the basic breeds in cattle, sheep and pigs by a process of careful mating. Darwin was particularly interested in pigeon breeding and was convinced, as were other pigeon fanciers at the time, that all the different types of pigeons had been derived from a common ancestor. 'Great as are the differences between the breeds of the pigeon, I am fully convinced that the common opinion of naturalists is correct, namely, that all are derived from the rock pigeon, *Columba livia* . . .' he wrote in *The Origin of Species*.

As you can see in figure 2.21 some of the varieties of pigeons, although desired by breeders, would be unlikely to survive in natural conditions. The same thing is true of many other species of animals which have been selected for particular characteristics. The enormous range of present day forms of dogs is evidence of selection stretching back over centuries. In some the emphasis in breeding has been to change the relative proportions of parts of the body as in the short-legged breeds like dachshunds bred for badger digging, or the heavy-headed bulldogs used for bull baiting – a 'sport' which was abolished in 1838. In others, such as greyhounds, speed was the desired character. Nowadays many dogs are bred to conform to the standards of the Show Bench and the characteristics found desirable are not those which would either have been advantageous in the wild, or would have been what the original breeders wanted or needed. The principle, however, remains. It is possible to change the characteristics of a group of animals by selection. In domesticated animals such selection does not give rise to new species; the new varieties, although very different from each

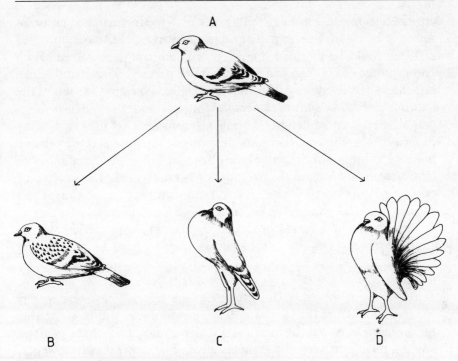

Figure 2.21   Variation in the domestic pigeon
A wild rock pigeon   B London pigeon 'blue chequer'
C pouter   D fantail
There are many varieties of the domestic pigeon, some of which are shown.
The variation results from mutation in the wild rock pigeon, *Columba livia*.
Pouters and fantails have been artificially selected by pigeon breeders.

other, can still interbreed and produce fertile offspring – even if there
may be some inherent mechanical difficulties because of size differences. A mating between a great dane and a dachshund, for example,
would bristle with technical difficulties.

Hybridization and polyploidy
In plant domestication, on the other hand, hundreds of new species
have been experimentally 'manufactured'. To do this two species with
the desired characteristics are crossed. However, there is a problem.
Species do not usually interbreed, and so even if some hybrid offspring
are produced they are likely to be sterile. The sterility *may* result from
the lack of homology between the chromosomes the offspring have

acquired from each of the parental species – the mismatched chromosomes will not be able to pair up during meiosis. If the chromosome number could be doubled then this problem could be solved. Each chromosome could then pair with a similar partner. The doubling can be achieved in many ways. Colchicine (an alkaloid from the roots of the autumn crocus, *Colchicum autumnale*), nitrous oxide or even heat shock will all stop spindle activity during metaphase. This means that the replicated chromosomes (chromatids) do not separate and the cells do not divide into two. So **polyploid** plants with double the number of chromosomes of each of the parent species are formed. The polyploids can breed amongst themselves, but not with either of the parent species, and therefore must be classed as a new species.

Sometimes the hybrids produced by crossing two species may be sterile even though the chromosomes may manage to pair during meiosis. In 1900 a new type of primrose was established at Kew Gardens, *Primula kewensis*. It was a hybrid between *Primula verticillata* and *Primula floribunda*. Unfortunately the hybrid was sterile. The chromosomes from the two parents were sufficiently similar to be able to pair in meiosis, but the mixed sets which could pass into the gametes seemed unable to keep the gametes alive. The new hybrid, however, could be maintained by vegetative propagation (which is a reason why polyploidy is much more common in plants than in animals). After many vegetatively reproduced generations a **tetraploid** shoot arose. From this one shoot a large number of tetraploid plants were raised. These had the desired characteristics, and also had larger leaves and flowers as a bonus. The tetraploid plants were also fertile and could breed amongst themselves. So the new species *Primula kewensis* was firmly established.

Polyploidy is a major **mode of speciation** with plants in natural conditions. It is thought that about half of all the species of flowering plants are polyploid, and one species of fern has 1260 chromosomes in its cells. So this is an area where artificial and natural selection can be seen to be rather similar processes. We return to consider the important topic of polyploidy in a later chapter. Not all plant breeding reaches the desired conclusion, however. In one case the aim was to cross a radish with a cabbage to produce a commercially useful plant with the good points of each species. Each parental species has eighteen chromosomes which produce nine pairs of homologues during meiosis. The hybrid produced was sterile as it had eighteen unpairable chromosomes. However polyploidy then occurred naturally to give a fertile plant with thirty-six chromosomes, and so things began to look good. The thirty-six chromosomes could make eighteen pairs during meiosis, and new offspring plants were grown. Unfortunately these had the roots of a cabbage and the leaves of a radish.

Artificial selection has also been carried out in the laboratory to test selection as an evolutionary force rather than to breed a new variety of plant or animal for a special purpose. Many different aspects of an organism's life have been changed by selection. Such controlled, experimental study of organisms such as bacteria and the fruit fly, *Drosophila*, is an important aspect of evolutionary research in many laboratories throughout the world. Some of these experiments are described in later chapters. However, although evolutionary change by selection can be studied in the laboratory it is more difficult to observe this in natural conditions. Nevertheless, many examples have been investigated over a relatively long period, and the conclusions drawn from observation have been supported by experimental work. The best documented cases are seen when the environment has changed in a particular way within a short time, as we describe in the next section.

The effect of a changing environment

The environment, at least in the industrialized nations, has been altered because of the various degrees and ways in which it has been polluted over the past few centuries. Industrial pollution has many effects. It kills most of the lichens which live on tree bark, leaving the tree trunks and branches bare to become soot covered. Soot and other types of industrial ash falls on leaves, reducing their efficiency as photosynthetic organs. Acidic gases such as sulphur dioxide which have been belched from factory chimneys into the atmosphere may be dissolved to fall as acid rain over areas which are a long way from the industrial regions. Solid wastes of all sorts have been dumped in large spoil heaps near mines or factories, destroying the vegetation which previously grew there. If the waste materials were not suitable to be dumped on land, they were pumped into rivers and lakes where they had a disastrous effect on the organisms which lived there. The catalogue of the spoliation of the environment seemed almost endless. At last man has accepted some responsibility for the sort of world he will hand on to his descendants, and a revolution in thinking about environmental matters is taking place.

If natural selection occurs and populations of organisms become more adapted to their local environment then such environmental changes which have resulted from pollution might be expected to produce corresponding changes in populations. There is a large amount of evidence that this has happened.

Industrial and other types of melanism

In many cases the changes have depended on the amount of camouflage an organism has against a particular background. Predators which hunt by sight kill a larger proportion of individuals which do not match their background well. Individuals which have the best camouflage are therefore likely to survive long enough to be able to reproduce and possibly pass on a similar pattern of camouflage to their offspring. So against different backgrounds, differently coloured individuals will be selected. Over eighty of the species of night-flying moths found in Britain have light and dark forms, the difference usually being due to the action of a single gene. The moths are good examples of **polymorphism** which was defined by E B Ford in 1940 as 'the occurrence together in the same habitat and at the same time of two or more distinct forms of a species in such proportions that the rarest of them cannot be maintained merely by recurrent mutation'.

The best known and most studied is the Peppered Moth, *Biston betularia*. In this the dark form, *carbonaria*, contains more of the dark pigment, melanin, because of the presence of a dominant allele; the light speckled form, typical, is the homozygous recessive, and a third group, *insularia*, result from other alleles at the same locus. Many investigators have studied changes in *Biston* populations, particularly Kettlewell in the early 1950s. An account of the research which has continued from then is given in most sixth form texts. Summarizing it briefly, it has been found that birds eat more of the moths which show up against the tree trunks. So in times of heavy pollution, there is an advantage to the moth in being dark coloured, and more of the light moths are eaten. This results in a change in the frequency of the light (and therefore also the dark) alleles in the population. If the level of pollution changes, as has happened in Britain since the passing of the Clean Air Act in 1956, there is a corresponding change in the frequency of the different alleles in the population. The changes are measurable even over a very short time span.

Many other melanic moths have been studied, not all of which are industrial melanics. Kettlewell, as well as investigating *Biston* populations in the 1950s, also collected moths from the Black Wood of

Rannoch. The wood, which is in Central Scotland, is one of the few remaining regions of the ancient pine forest which once covered much of Britain. The huge pines in the forest are only partly covered with lichen. Of the seven species of moths with melanic forms which Kettlewell found in the forest, he decided to concentrate on the Mottled Beauty, *Cleora repandata*. About 10 per cent of the 500 Mottled Beauty specimens he observed were dark. The moths spent the day on pine trunks, and here the light form was very well camouflaged. However, if Kettlewell noted where the moths rested at dawn and then revisited the tree later in the day, he found that a large number of moths had moved from their first resting place. During flight the dark form seemed to have an advantage. The dark forms became practically invisible when only eighteen metres away from an observer, whereas the light forms could be distinguished for nearly ninety-one metres. So both forms had their own advantage – the light forms were protected from predation when they were resting, and the dark forms were better protected during movement. Melanism in these moths is not the result of a recent mutation, and the two forms have been maintained in the population because each type has a particular advantage, depending on the circumstances. The Mottled Beauty is also found in urban areas. In the London area and also in Bradford there are melanic specimens which seem very similar to those found in the Black Wood. About 30 per cent of the London Mottled Beauty specimens are melanic, but in Bradford the melanic form has apparently completely replaced the typical form.

Insects other than moths may also show industrial melanism. The common Two-spot Ladybird, *Adalia bipunctata*, has dark forms in

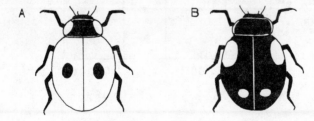

Figure 2.22 Some polymorphs of *Adalia bipunctata*, the two-spot ladybird
A red form with one black spot on each side
B black form with two red spots on each side

which the wing covers are black, with a fixed number of red spots; it also has a typical form with red wing covers with two black spots, one on each side. These various forms are mainly controlled by a single gene which has a series of multiple alleles. The most dominant allele gives black wing cases with one red spot on each side; the least dominant allele gives the typical red form. Between these there are two other forms – black with two red spots on each side, and black with three red spots on each side. So the amount of red increases from the most dominant to the least dominant. The red forms occur most commonly in rural areas, whereas about 97 per cent of the specimens around Liverpool and Glasgow are the black forms.

However, the story must be more complicated than that of the

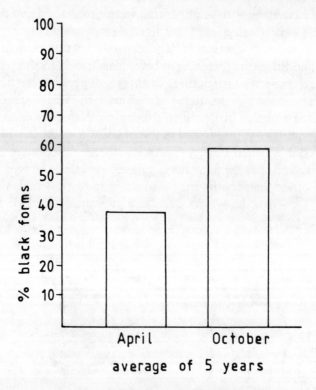

Figure 2.23   Average percentage of black forms of *Adalia bipunctata* in Spring and Autumn (taken over five non-consecutive years)

Black forms are rarer than red in the Spring as they do not survive hibernation as well, but they do better than red during the summer.

Peppered Moth. Ladybirds are protected against predation by a powerfully unpleasant taste and scent. Only Redstarts regularly eat ladybirds, and their effect on numbers is not important. Timofeeff-Ressovsky counted the numbers of black and red forms in a locality near Berlin over five (not consecutive) years. He found that the red forms were commoner in the spring, and black forms commoner in autumn. The red form survives the winter hibernation better than the black, but the black does better in the summer. Marriner found, by laboratory experiments, that black ladybirds flourished in temperatures which were high enough to kill the red forms.

The story became even more complicated when Creed showed that the type of industrial pollution affected the distribution of these ladybirds in Britain. There were very few of the black forms found in South Wales, although this was a highly industrialized region. However, the type of coal burnt in South Wales did not produce such a heavy smoke cover in the air as in the northern industrial areas. If such a heavy smoke cover substantially reduces the amount of sunshine reaching the ground, then the black forms are at an advantage – their dark colour allows them to absorb more of the available radiation. So the high frequency of black forms in some industrially polluted regions is not, as in the Peppered Moth, because they are less frequently eaten than the lighter forms but because they can combat the lack of sunshine better. The black forms have decreased in frequency in the Birmingham area since the passing of the Clean Air Act. But how can we explain a frequency of 63 per cent black forms in Harrogate, and 75 per cent black in Hexham – both being in largely rural areas? As in the case of the Mottled Beauty, it looks as if an old variation may be subjected to a different type and strength of selection in different local circumstances.

Effects of pollution on plant growth

Plant growth is affected by industrial pollution in many ways. Atmospheric smoke can reduce the amount of light which falls on leaves, and therefore cut photosynthesis rates. The clogging of the leaf surfaces with solid pollutants can have the same effect. The presence of the ions of heavy metals such as copper, zinc and lead in the soil has a disastrous effect on root growth in many plants. However, there are variations in the **tolerance** shown to the metals by various plants. In the common grass *Agrostis tenuis* some plants are tolerant of heavy pollution, and all grades of tolerance are shown between these plants and those which are highly susceptible. Bradshaw has investigated, over many years, the way in which the grass has grown in soils of different degrees of pollution. In populations growing on unpolluted soil there are one or two tolerant plants per thousand susceptible plants. However, no susceptible plant can grow in polluted soil, and so there is very strong selection in favour of the tolerant plants there. The grass is wind-pollinated and so the tolerant plants may receive pollen from plants growing in unpolluted soil, but selection will eliminate all but the tolerant of their offspring. Selection acts in the other direction in unpolluted soil, although the tolerant plants are not eliminated completely. Tolerant plants cannot grow well in the crowded conditions found in the normal, unpolluted pasture. Old spoil heaps are now being made less unsightly by being seeded with tolerant varieties of grasses.

A similar rapid selection in favour of metal tolerant plants has caused rather large economic problems for ship owners. In this case, as you will realize, it is not grasses which are involved. Until fairly recently ships needed to have a period out of service during which they had their bottoms scraped to get rid of all the barnacles and other marine organisms which had settled there. Apart from the cost of the scraping, there was the additional cost of the loss of the ship's earnings during the time. So antifouling paint was developed and this has reduced the barnacle growth to a minor problem. Antifouling paint works by releasing toxic compounds (usually copper but also some other metals), and this has stopped the attachment of barnacles. However, some species of the green alga *Ectocarpus* have continued to grow on the hulls even after antifouling treatment. So the ships still need to be scraped. The tolerant strains of algae grow so well on the hulls that they can add extra drag to the ship's movement through the water, and this means that extra fuel must be used during a trip. *Ectocarpus silculosus* taken from the bottoms of ships has been shown to be tolerant to ten times the concentration of copper than that which *Ectocarpus* plants growing on a normal rocky shore can stand. So after a brief time in which the antifouling paints were of great value, their effects have been reduced by the rapid selection of plants which can survive in the high local mineral concentrations they produce.

Many other examples of rapid change in the gene frequency in populations as a result of environmental change could be given, and some of these we discuss in later chapters. However, these few examples given here provide evidence that evolutionary change can, and does, occur. In the next chapter we consider how the genetic variations, which form the raw material for evolution, can arise.

SUMMARY

Because of the enormous time span since the beginning of life on earth no one can be certain of what has happened. However, there are various lines of investigation which can provide **evidence** to give us some idea of the changes which have occurred in organisms during this time. The evidence must, of course, be interpreted – a **fossil**, for example, is meaningless unless one has some idea of **when** it was fossilized and how it might **relate** to other fossils or present day organisms. Similarly, the present day **distribution of organisms** can be understood if it is related to the **past movement**, either of the organisms themselves or of the land on which they lived. **Geographically isolated** organisms have diverged from their **common ancestors** as a result of **adaptation** to the particular conditions in which they live. Present day organisms show many **anatomical** and **physiological characters** which suggest their relationships with each other, and these can also be compared with characters shown in fossil species. All organisms use the same fundamental **biochemical reactions** and **molecules** during their metabolism. Although it is impossible to check the biochemistry of long-dead or extinct organisms, there is no reason to believe that it differed in any fundamental way from that which is found in organisms today. However, there are minor differences which help to sort out how closely organisms are related to each other and which might give some idea of when they diverged from common ancestors. This **natural divergence** resulting from such evolutionary forces as **natural selection** can also be **artificially** brought about by the deliberate choice of plant and animal breeders and **new species** may be produced in a fraction of the time needed in nature. Similarly, as organisms are adapted to the environment, a **rapid change** in a particular environment might be accompanied by a related rapid change in the characters of the organisms which live there.

FURTHER READING

Armstrong, Patrick, 'The importance of being insular', *New Scientist*, vol. 84, no. 1194 (14 February 1980) p. 494.

Barrington, E J W, (ed.), *Hormones and Evolution* (Academic Press, London and New York, 1979).

Bell, Graham, *The Masterpiece of Nature. The Evolution and Genetics of Sexuality* (Croom Helm Applied Biology Series, 1982).

Blundell, T L, Humbel, R E, 'Hormone families: pancreatic hormones and homologous growth factors', *Nature*, vol. 287, no. 5785 (30 October 1980) pp. 781–787.

Bradshaw, A D, 'Plant evolution in extreme environments', in Creed, R, (ed.) *Ecological Genetics and Evolution* (Blackwell, 1971) pp. 20–49.

Brown, R G B, 'Species isolation between the Herring gull, *Larus argentatus*, and Lesser Blackbacked gull, *Larus fuscus*', *Ibis*, vol. 109 (1967) pp. 310–317.

Creed, R, (ed.), *Ecological Genetics and Evolution* (Blackwell, 1971).

Dayhoff, Margaret Oakley, 'Computer analysis of protein evolution', *Scientific American* (July 1969).

Diamond, Jared M, May, Robert M, 'Island biogeography and the design of natural resources', in May, Robert M (ed.) *Theoretical Ecology* (2nd ed.) (Blackwell Scientific Publications, 1981).

Dickerson, Richard E, 'Chemical evolution and the origin of life', *Scientific American* (September 1978).

Dietz, Robert S, Holden, John C, 'The break-up of Pangaea', *Scientific American* (October 1970).

Dollar, A T J, Guest, J E, 'Iceland's new island volcano', *New Scientist*, vol. 20, no. 368 (5 December 1963) pp. 591–593.

Endler, J A, 'Gene flow and population differentiation', *Science N.Y.*, vol. 179 (1973) pp. 243–250.

Fitch, Walter M, Margoliash, Emanuel, 'Construction of Phylogenetic Trees', *Science*, vol. 155, no. 3760 (1967) pp. 279–284.

Fitch, Walter M, Margoliash, Emanuel, 'The usefulness of amino acid and nucleotide sequences in evolutionary studies', in *Evolutionary Biology*, vol. 4, Dobzhansky, Theodosius, Hecht, Max H, Steere, William C, (eds.) (Appleton – Century Crofts, 1970).

Gruerrier-Takada, Cecilia, Gardiner, Kathleen, Marsh, Terry, Pace, Norman, Altman, Sidney, 'The RNA moiety of ribonuclease P is the catalytic subunit of the enzyme', *Cell*, vol. 35 (1983) pp. 849–857.

Johnston, R F and Selander, R K, 'House sparrows; rapid evolution of races in North America', *Science*, vol. 144 (1964) pp. 548–550.

Kemp, Tom, 'Fossils complete the record', *New Scientist*, vol. 93, no. 1291 (4 February 1982).

Kettlewell, H B D, *The Evolution of Melanism* (Clarendon Press, 1973).

Lack, David, *Darwin's Finches* (Cambridge University Press, 1947. Reprinted Peter Smith, Gloucester, Mass., 1968).

Lack, David, *Population Studies of Birds* (Clarendon Press, 1966).

Lewis, H E, 'The Tristan Islanders: a medical study of isolation', *New Scientist*, vol. 20, no. 370 (19 December 1963).

May, Robert M, 'The evolution of ecological systems', *Scientific American* (September 1978).

Ridley, Mark, 'Evolution and gaps in the fossil record', *Nature*, vol. 286, no. 5772 (31 July 1980) pp. 444–445.

Schopf, J William, 'The evolution of the earliest cells', *Scientific American* (September 1978).

Sheppard, P M, *Natural Selection and Heredity* (Hutchinson University Library, 1975).

Stanley, Steven M, *The New Evolutionary Timetable* (Basic Books, Inc., N.Y., 1981).

Valentine, James M, 'The evolution of multicellular plants and animals', *Scientific American* (September 1978).

3

The raw material
for evolution

MENDEL'S WORK

Mendel was perhaps fortunate in choosing pea plants and also the particular characters he observed in his classic studies on inheritance. Pea plants show a number of clear cut phenotypic differences which we now know are controlled by single genes. Seven pairs of characters formed the main part of his research (in each case the dominant allele is given first): round cotyledons/wrinkled cotyledons; yellow cotyledons/ green cotyledons; greyish purple testa/white testa; inflated pods/ constricted pods; green pods/yellow pods; axial flowers/terminal flowers; tall plants/short plants. Pea plants are also self-fertilized and so phenotypic differences between lines of plants can be maintained.

Mendel was not the first investigator to try to discover how heredity worked, but earlier plant and animal breeders had investigated traits which seemed more useful, such as body mass in animals. Unfortunately such traits are very often determined by many genes working together (**polygene systems**). The environment may also have a great effect on the phenotype. The characters Mendel chose to investigate gave consistent results in the fairly constant environment of the garden of the Augustinian monastery at Brunn in Moravia. Mendel's knowledge of the dominant characters came from careful observation and study. His mathematical analysis of a **dihybrid cross** would not have been straightforward, however, if he had by chance chosen characters which were controlled by **linked genes**. The very variable numbers of parental and **recombinant forms** which result from crossing-over in meiosis would have made it impossible for Mendel to derive a statistical basis for his theory. But he chose genes which were on different chromosomes and which therefore showed independent assortment.

The results he obtained from a very large number of crosses supported the hypotheses he had suggested for the mechanism of inheritance. In fact, some of the results he claimed seem to be statistically too perfect. Nevertheless, he demonstrated that the genetic factors (as he called them) were **particulate** and maintained their **integrity** through many generations. Even if one of the two parental types seemed to have disappeared in the offspring produced by crossing two different homozygotes, it could reappear in the second generation.

The development of the science of genetics after the rediscovery of Mendel's work in 1900 brought to light many examples which initially did not seem to fit the Mendelian rules. More had been discovered about chromosomes, and the way they behaved during meiosis, in the intervening years. The phenomenon of **linkage**, and the way in which new combinations of alleles could be made by genetic **crossing-over** explained some of the anomalous results. The effect of a variable environment on the functioning of a gene was recognized and the concepts of **genotype** and **phenotype** were established. It became clear that many phenotypic effects were controlled by more than one gene. Sometimes the functioning of one gene had an inhibiting effect on the functioning of another. Sometimes two genes acting together produced a new character.

GENE INTERACTION

Such gene interactions can be explained by considering the role of enzymes in the metabolism of the organism. Beadle and Tatum proposed a **one gene–one enzyme** (now considered as **one gene–one polypeptide**) theory as a result of their work, from the 1930s onward, on the red bread mould, *Neurospora crassa*. However, their work was predated by the ideas put forward by Sir Archibald Garrod in his book *Inborn Errors of Metabolism* in 1908. He had been studying a hereditary disease called alkaptonuria in which a striking symptom is that the urine turns black if it is exposed to air, because it contains homogentisic acid. He reasoned that an enzyme was missing which would normally convert the homogentisic acid into the acetoacetic acid which is usually found in the urine. Homogentisic acid is a derivative of the amino acid tyrosine which the patient ingests in his protein food. Garrod showed that the amount of homogentisic acid in the urine of these patients was increased if he increased the amount of tyrosine they took in. From this he brilliantly suggested that many other hereditary diseases are the result of a loss of an enzyme which controls some part of a metabolic chain reaction, a suggestion which has been proved correct by subsequent work in physiological genetics. See figure 3.1.

The metabolism of a cell is a highly efficient, closely linked series of chemical reactions in which the products of one reaction become the substrates of another. The lack of an enzyme at some point, or a change in the cellular environment such as altered local temperature or pH, can therefore affect the overall process. Products which are blocked from becoming substrates may accumulate. They may stay within the cell or be carried away to other parts of the organism.

In *Drosophila melanogaster* the normal wild-type eye colour is dark red but mutations can produce eyes which are white, different shades of pink, bright red or brown. Two types of pigment are involved in the

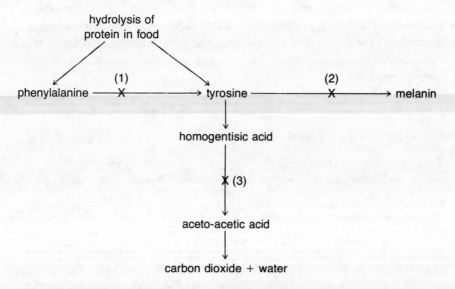

X   – chemical block because of lack of the appropriate enzyme
(1) – lack of hydroxylase (homozygous *pp*) – phenylalanine accumulates, leading to phenyl-
     ketonuria.
(2) – lack of tyrosinase (homozygous *cc*) – no melanin can be formed, leading to albinism.
(3) – lack of homogentisate oxidase (homozygous *aa*) – homogentisic acid accumulates,
     leading to alkaptonuria.

Figure 3.1   The way in which mutation may affect the metabolism of the amino acids phenylalanine and tyrosine, leading to human abnormalities

normal dark red eye colour – pterins which are bright red and ommo-
chromes which are brown. A mutation which prevents the production
of pterins will lead to brown eyes and one which prevents ommochrome
production will lead to bright red eyes. White eyes result from an
inability to make either pigment. It is known that mutations in at least
twenty-six genes can affect eye colour and so the production of each
pigment must be controlled by several different genes. Each stage in the
sequence of chemical reactions needed for pigment production is
enzyme controlled and if the enzyme is missing the sequence will stop.
If this happens then the substrate for the next stage will accumulate and
give rise to the abnormal colour. Sometimes the eyes are white not
because of a mutated inability to *make* the pigments but because of an
alteration in the protein which normally acts to carry the pigment from
its place of origin to its final position. The protein is useless as a carrier
molecule and the pigment diffuses away. Many different genes are
therefore involved in producing what appears to be a simple phenotypic
trait.

In some cases genes have a cumulative effect, as in human skin
colour. A possible model to explain the large amount of variation which
occurs suggests that five genes are involved. Each gene has two alleles,
only one of which leads to melanin production. The very darkest skin
colour would result from the presence of ten alleles for melanin production A★A★B★B★C★C★D★D★E★E★ – and the very lightest from the
converse situation of ten non-melanic alleles – AABBCCDDEE. Various crossings would produce a normal distribution ranging from zero to
ten pigment alleles, with the intermediate genotypes being most common. The large numbers of possible genotypes result from the
reshuffling which occurs at meiosis – five pairs of homologous chromosomes can produce 2^5 possible arrangements within gametes. As any
gamete can theoretically join to any other gamete at fertilization, provided they are of the right sex, the range of possible variation becomes
very large.

Sometimes a single mutation can produce widespread and seemingly
separate phenotypic effects. Haemoglobin, the oxygen carrier in the red
blood cells of all vertebrates, consists of four polypeptide chains of two
types, alpha globin and beta globin, and iron containing porphyrin
rings, haem. The genes controlling the formation of the globins are on
different chromosomes. In human haemoglobin the alpha chains are of
141 amino acids and the beta chains of 146 amino acids. Over 100
different human haemoglobins have been discovered, most of them

Alpha chain variants

Amino acid number	Amino acid substitution	Name
16	lysine to glutamic acid	I
23	glutamic acid to glutamine	Memphis
30	glutamic acid to glutamine	G Chinese
92	arginine to glutamine	J Cape Town
92	arginine to leucine	Chesapeake
102	serine to arginine	Manitoba

Beta chain variants

Amino acid number	Amino acid substitution	Name
6	glutamic acid to valine	S
58	proline to arginine	Dhofar
63	histidine to arginine	Zurich
88	leucine to proline	Santa Ana
132	lysine to glutamine	K Woolwich
145	tyrosine to histidine	Rainier

Figure 3.2 List of some human haemoglobin substitutions *Data from Lehmann and Carrell.*

differing from each other in one or two amino acids in an alpha or beta chain. Some of them do not seem to produce any clinical symptoms, such as haemoglobin G and haemoglobin Dhofar. Haemoglobin Zurich also seems to produce little effect in the person, as long as the individual is not given sulphanilamide. In these circumstances anaemia results. The most important example of illness arising from a changed haemoglobin is that of sickle cell anaemia. A substitution of adenine for thymine in one of the genes coding for the beta chains leads to the incorporation of valine instead of glutamic acid in the sixth position from the amino end of the chain. The abnormal haemoglobin, haemoglobin S, containing this precipitates out at low oxygen concentration. This gives the red cells a sickle shape, from which the illness is named. The cells tend to clog small blood vessels. From the primary cause of a single mutation a large variety of symptoms can arise.

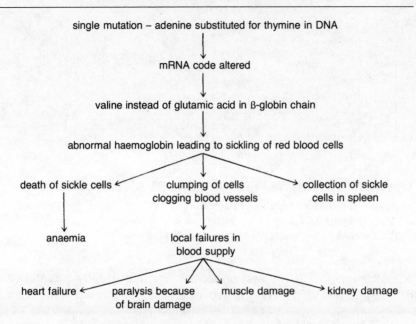

Figure 3.3   Some results from a single DNA change in the gene coding for one of the polypeptides which make human haemoglobin

GENETIC CODE

The original source of genetic variation is the process of mutation, a process in which the DNA of an organism is altered either in its constituents or in its quantity. Gene or point mutations involve the substitution of one base for another in the DNA molecule. This produces a change in the coding of the DNA and possibly a change in the amino acid sequence of the protein into which it may be transcribed and translated.

A base substitution does not always lead to a changed amino acid, however. As the DNA base order is transcribed onto mRNA in groups of three bases, and there are four bases available there are 4^3 (sixty-four) possible permutations or codons. This is a much higher number than would seem to be needed to code for the twenty amino acids which make up proteins. The DNA code is **redundant** – when transcribed onto mRNA there are several **codons** which specify the same amino acid. As you can see in figure 3.5, the third base in each codon may usually be altered without changing the amino acid specified. However,

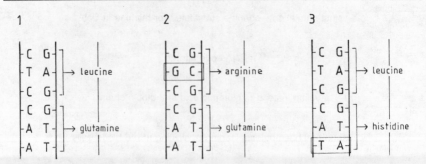

Figure 3.4   The effect of a mutation in the DNA on the protein formed
1 original DNA and section of protein
2 and 3 mutant DNA and changed protein
G guanine   C cytosine   A adenine   T thymine

a point mutation at the first or second position usually means that a different amino acid will be incorporated into the protein. Amino acids may be grouped together according to their chemical properties. Some are acidic, some basic and some neutral, whereas others are aromatic or contain sulphur. A mutation which results in the substitution of one amino acid for another in the same chemical group has rather slight effects on the structure and function of the protein which is made; one which changes an amino acid for another in a different chemical group is likely to have more drastic effects.

The genetic code shown in figure 3.5 was elucidated by biochemists inspired by a momentous discovery by Nirenberg in 1961. Nirenberg had found that an artificial RNA which consisted entirely of uracil **nucleotides** ('poly U') would, if added to a mix of polypeptide making enzymes from *Escherichia coli*, the colon bacillus, and a radioactively-labelled collection of all the twenty amino acids, instigate the production of a polypeptide which contained only phenylalanine. If the artificial RNA used had more complex sequences such as UUU;GUG;UUU;GUG . . . then the polypeptide produced contained the amino acid valine instead of phenylalanine at the expected places – phenylalanine;valine;phenylalanine;valine and so on. Within a few years the whole **genetic dictionary** had been deciphered.

The way in which information passed in cells from the code in the base arrangement of the DNA to the amino acid sequence of the finished protein was now clear and was defined in the '**central dogma**':

DNA ⟶ RNA ⟶ protein

Figure 3.5 The genetic code in the language of mRNA. The code is universal – all organisms use the same set of RNA codons to specify all the twenty amino acids. AUG codes for a 'start' signal, but there are 3 'stop' signals – UAA, UAG and UGA

First RNA nucleotide base	Second RNA nucleotide base				Third RNA nucleotide base
	U	C	A	G	
uracil(U)	phenylalanine	serine	tyrosine	cysteine	U
	phenylalanine [1]	serine	tyrosine [1]	cysteine [4]	C
	leucine	serine	STOP	STOP [5]	A
	leucine	serine	STOP [5]	tryptophan [1]	G
cytosine(C)	leucine	proline	histidine	arginine	U
	leucine	proline	histidine [2]	arginine	C
	leucine	proline	glutamine	arginine	A
	leucine	proline	glutamine	arginine [2]	G
adenine(A)	isoleucine	threonine	asparagine	serine	U
	isoleucine	threonine	asparagine	serine	C
	isoleucine	threonine	lysine	arginine	A
	methionine/START [4]	threonine	lysine [2]	arginine [2]	G
guanine(G)	valine	alanine	aspartic acid	glycine	U
	valine	alanine	aspartic acid	glycine	C
	valine	alanine	glutamic acid	glycine	A
	valine	alanine	glutamic acid [3]	glycine	G

[1] aromatic amino acids
[2] basic amino acids
[3] acidic amino acids
[4] sulphur containing amino acids
[5] stop signals

Sections without numbers are neutral amino acids.

The DNA contains the code and the protein is made according to the instructions in the code. The 'central dogma' may also be shown as follows:

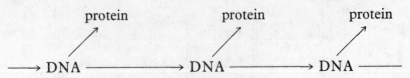

Such a concept is very similar to the **germ plasm theory** proposed by Weismann in the Nineteenth Century. He regarded the fertilized egg as being the starting point for two independent processes:

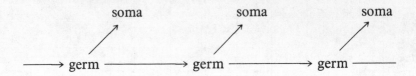

The diagram shows three generations of an organism. In each generation one of the two processes leads to the production of the 'soma' or individual body which can be acted on by the environment, and which is mortal; the other process leads to a very early differentiation of future 'germ line' cells which will give rise to gametes and so produce the next generation. The 'soma' is produced from the 'germ line' and the reverse process does not occur. This means that Lamarck's idea of the inheritance of acquired characters must be rejected. Changes in the body which have been induced by the environment cannot be passed on to the next generation unless the DNA in the gamete forming cells has also been altered in the 'right' way. Mutations which affect the DNA of some body cells, **somatic mutations**, may show their effects in the particular organism but not in subsequent generations, whereas mutations in the cells of the 'germ line' will be found in all the cells of the offspring.

ANALYSIS OF PROTEINS

An analysis of the proteins of an organism will therefore show changes in its DNA. The proteins may be released by the homogenization of some tissue, and can be analysed by the process of **electrophoresis**.

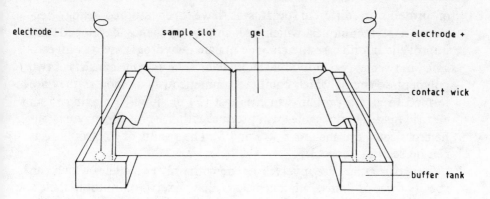

Figure 3.6 Gel electrophoresis

A tissue sample from each of the organisms being studied is homogenized to release the proteins. The proteins are placed in the sample slot on the gel and the current switched on for some hours. The gel is then treated with a solution containing the substrate for the enzyme being studied, and a particular salt. The enzyme converts the substrate into a product which is stained by the salt to give coloured bands.

Each amino acid, polypeptide or protein molecule has one end positively and the other end negatively charged, and forms a zwitterion. This, on its own, would give electrical neutrality to the molecule but the side chains which determine the properties of individual amino acids may also be charged. There would then be a net overall charge on the molecule which might be positive or negative. Proteins are therefore likely to differ electrically from each other depending on the amino acids from which they are made. In 1966 Lewontin and Hubby used this principle in the first extensive analysis of protein polymorphisms in *Drosophila pseudoobscura*. Extracts of individual flies were subjected to gel electrophoresis.

The tissue extracts were placed separately in the sample slot of a gel made of polyacrylamide, starch, cellulose, agar or some other substance which made a homogeneous base. Contact wicks passed from tanks containing electrodes in a suitable buffer solution to the gel and the current was switched on for several hours. The proteins migrated according to their net electrical charge and also their molecular size. Their final positions were found by staining.

Eighteen different **genetic loci** in *Drosophila pseudoobscura* were investigated in the first series of experiments. Since then a large number of proteins in different organisms have been studied using electrophoretic techniques which are constantly being developed and improved. In one case the enzyme malate dehydrogenase was investigated in twenty-two individual *Drosophila*. The enzyme consists of two polypeptide chains which combine spontaneously as soon as they are formed to give a protein which is fast (F) or slow (S) moving in an electric field. Homozygotes will produce either FF or SS enzymes but heterozygotes can make FF, SS or FS. The results of electrophoresis can be seen in figure 3.7.

This particular case involves a gene with only two alleles which can specify three proteins. Other genes may have a series of multiple alleles, and the large number of proteins which can be produced makes the electrophoretic analysis much more complicated. The enzyme xanthine dehydrogenase in *Drosophila pseudoobscura* was originally investigated by simple electrophoresis on a 5 per cent polyacrylamide gel at pH 8.9, and six alleles of the gene were revealed. The experiment was then repeated using apparently homogeneous inbred flies but on gels of different concentrations and at different pH values. The present count of alleles at the locus stands at thirty-seven. As these thirty-seven alleles were detected in only 146 flies, it is likely that the final number of alleles at this locus will be even higher.

Electrophoresis, of course, underestimates the number of mutations which might have occurred. A change in the third base of a codon

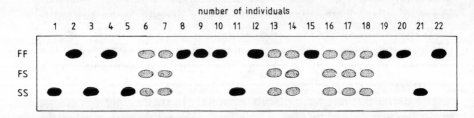

Figure 3.7 Electrophoretic gel stained for malate dehydrogenase from 22 *Drosophila* individuals

Tissue samples from the 22 flies were regularly spaced in the sample slot and migrated under electrophoresis to give the pattern shown. The enzyme is made from 2 polypeptide chains which join spontaneously as soon as they are produced. Homozygotes such as flies 1 and 2 produce enzymes which are slow or fast moving, SS or FF. Heterozygotes produce 3 types of enzymes – SS, FS and FF – like fly 6. *Modified from photograph in Ayala.*

usually has no effect on the coding for an amino acid (look back to figure 3.5), and so the protein will be unchanged. Even if the coding is changed by mutation so that a different amino acid is incorporated into the protein, it will not be detectable by electrophoresis unless it has altered the overall charge on the protein molecule.

MUTATION – TYPES AND RATES

Electrophoresis *can* show many of the changes in amino acid sequence which result from mutation. Gene mutation produces the raw material for the code change; the genes are then scrambled by the processes of recombination (which provides new combinations of alleles on the same chromosome) and random assortment (which gives new combinations of chromosomes in the gametes).

Reproduction in asexual organisms is not brought about by meiosis and so these scrambling processes are not available to them. However, bacteria can reproduce so rapidly that new mutations may be incorporated in a very short time. Bacteria also have some other ways of changing their genetic material. **Conjugation** between two bacterial cells can involve the passage of a small length of DNA from one cell to another. Bacteria can also, in certain circumstances, pick up DNA (which might have come from dead bacterial cells) from the surrounding medium in a process of **transformation**. The particular sorts of viruses which attack bacteria (bacteriophages) may also transfer some bacterial DNA from one cell to another – this is called **transduction**. So although bacteria are basically asexual, they do have methods of getting a new genetic mix.

Up to about twenty years ago it was thought that organisms were likely to be homozygous for the wild type allele at nearly all their gene loci. Mutant alleles, which are usually deleterious in their effects, might be constantly introduced by chance giving a few heterozygous loci, but were just as constantly removed by selection. However, this theory could not account for the large amount of variation shown by electrophoresis. Now each gene is thought to have a wide array of possible alleles, and an organism is likely to be heterozygous at many of its loci. This means that there is no 'normal' or ideal genotype. There will, however, be a range of genotypes which are satisfactorily adapted to the types of environment usually encountered by the population.

As can be seen in figure 3.8 invertebrates generally show more heterozygosity than vertebrates. A survey of a population of *Euphausia superba* (the small crustacean named krill which is an important source of food for whales) showed that although fifteen of the thirty-six loci in the 126 sampled krill individuals were invariable, the other twenty-one loci had two, three or four possible alleles. Man has a heterozygosity of 6.7 per cent – close to the average for vertebrates. Tests suggest that there are 10 000 gene loci in man, although the actual amount of DNA present suggests a higher figure. Even 10 000 distinct loci would mean that each individual is likely to be heterozygous at 670 loci (6.7 per cent of 10 000). This amount of heterozygosity, with the effects of crossing-over, would give each person the potential of producing 2^{670} different gametes. Such a vast number makes it almost impossible for any two people (apart from identical twins or other **clones**) to be genetically identical (which is unfortunate for the writers of fiction,

Organism	Heterozygous loci per individual (%)
Invertebrates	
Drosophila	15
Wasps	6.3
Other insects	15
Marine invertebrates	12
Land snails	15
Vertebrates	
Fishes	7.8
Amphibians	8.3
Reptiles	4.7
Birds	4.3
Mammals	5.2
Average values	
Invertebrates	13.4
Vertebrates	6.2
Plants	17.3

Figure 3.8   Amount of genetic variation in natural populations, as estimated by gel electrophoresis

In general invertebrates show more variation than vertebrates. Only a few plant species have been studied but they seem to show a large amount of genetic variation. *Data from Ayala*.

where 'doubles' or 'doppelgangers' seem to crop up at a rather high frequency). So most of the variation in population arises from the reshuffling process rather than from new mutation.

Of course, single base mutations and the subsequent effects of recombination and independent assortment are not the only ways in which genetic variation may be produced. There may be changes in the actual amount of DNA present by means of **deletions** or **additions**. These may be at the **single base level** (thus possibly altering the coding pattern for the rest of the gene) or involve the loss or addition of **large pieces of chromosome** made up of many genes. Whole sets of chromosomes may be duplicated if a spindle does not form during cell division, thereby leaving the replicated sets of chromosomes in an undivided **polyploid** cell. Sometimes one chromosome only out of a cell is affected. The two chromosomes show non-disjunction and do not separate. This produces a condition of **aneuploidy** in which one daughter cell has one chromosome too many (**trisomy**) and the other cell has one too few (**monosomy**). All the chromosomes in a set are important to the cell and a polyploid cell, because it has the same number of each chromosome, can usually function as well as, or even better than, a normal diploid cell. Aneuploidy, on the other hand, produces a cell with an unbalanced number of chromosomes and this usually results in a less fit organism.

The process of cell division, whether mitosis or meiosis, is so complicated a manoeuvre that it is not surprising that things sometimes go wrong. Think of what is involved. First of all the DNA must replicate during interphase to make the chromatids. Nucleotides must be available in suitable quantities and be arranged by enzyme action in the correct order, complementary to each of the sides of the unzipped DNA molecule. The ·nuclear membrane must break down, to be reformed later. A spindle must form from microtubules, which themselves must be assembled from protein units. The chromatids must be able to attach to the spindle fibres by their centromeres in the correct alignment, and be moved apart by changes in the chemical configuration of the fibres. A fault in DNA replication; an alteration in spindle formation; a lack of any of a large number of enzymes; a break in a chromosome to give a section without a centromere which therefore has no attachment point; all of these could produce drastic effects. These are chance events; each could happen because of a change in the internal environment within the cell or because there has been a change in the external environment.

Mutagens

Radiation

A change in the external environment can also have an effect on mutation rates. Organisms are constantly subjected to various types of radiation. The **electromagnetic spectrum** extends from long radio waves which may have wavelengths of several kilometres down to cosmic rays with wavelengths of 10^{-16} metres.

The amount of energy contained in the radiation becomes larger as the wavelength becomes shorter, and above a certain energy level the rays can penetrate tissues and cells.

Ultraviolet radiation can penetrate tissues less well than higher energy rays but is readily absorbed by the bases in nucleic acids, particularly by cytosine and thymine. Cytosine may be hydrolysed to a product which causes mispairing. Thymine may link to an adjacent thymine to form a dimer which acts like a blocked zip fastener and so prevents replication. It is possible for the zipper to be unblocked again if the dimer has been formed on only one side of the DNA molecule.

Ultraviolet radiation is a normal part of the sunlight which reaches the earth's surface, and most organisms have some protection from its mutagenic effect. Melanin production in human skin may be temporarily increased during the process of sun-tanning, and selection for particular melanin levels is seen in races living in different geographical regions. The repair system allows much of the damage which might occur in spite of the shielding effect of melanin to be quickly eliminated so that it does not become permanent. People with the inherited disease *Xeroderma pigmentosum* produce too little of the enzymes needed to carry out the repairs and the ultraviolet induced damage in their skin cells may lead to cancerous conditions.

Radiation of wavelength shorter than that found in ultraviolet rays is known as **ionizing radiation**. The energy level is so high that electrons in the irradiated atoms may be knocked out of orbit, thus producing a positively charged ion. Ions, and the molecules which contain them, are chemically much more reactive than the original neutral atoms. The electrons which are lost by one atom in the process move off at high speed and are taken up by another atom which therefore becomes negatively charged. So a chain of ion pairs (one positive and one negative) is found along the path of the high energy radiation. The ion pairs have a short existence, reverting rapidly to neutrality, but it is during this reversion that chemical reactions can lead to mutation. One

frequency (Hz)	type of radiation		wavelength (m)
10^{24}	cosmic rays		10^{-16}
10^{23}	gamma rays		10^{-15}
10^{22}			10^{-14}
10^{21}			10^{-13}
10^{20}			10^{-12}
10^{19}	X rays		10^{-11}
10^{18}			10^{-10}
10^{17}			10^{-9}
10^{16}	ultra-violet radiation		10^{-8}
10^{15}			10^{-7}
10^{14}			10^{-6}
10^{13}	infra-red radiation		10^{-5}
10^{12}			10^{-4}
10^{11}	microwaves	EHF	10^{-3}
10^{10}		SHF / radar	10^{-2}
10^{9}		UHF	10^{-1}
10^{8}	television	VHF	1
10^{7}		HF	10
10^{6}	AM radio	MF	10^{2}
10^{5}		LF	10^{3}
10^{4}		VLF	10^{4}
			10^{5}

Figure 3.9 The electromagnetic spectrum
The shaded area is visible light.

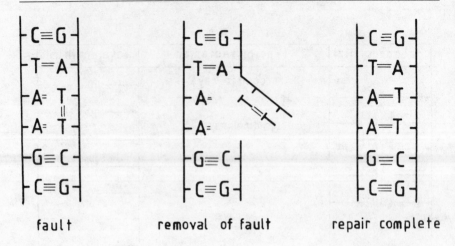

fault   removal of fault   repair complete

Figure 3.10   Repair mechanism for UV damage to one strand of DNA

Two adjacent thymines have become linked together instead of to their complementary bases. The defective section is removed by means of the enzyme endonuclease which nicks the sugar-phosphate backbone of the DNA. New nucleotides are then inserted to fill the gap and the enzyme polynucleotide ligase completes the repair.

result is breakage of the sugar-phosphate backbone of the DNA molecule. If the strand breaks at two closely spaced points there could be a single base deletion. Chromosomal structural changes may result from breaks which are more widely separated. The subsequent effects of irradiation depend on the type and position of the irradiated cells. Any mutations caused will not be maintained in the population unless they occur in the gamete producing cells.

Organisms are subjected to low levels of ionizing radiation from cosmic rays and from radioactive materials in the earth's surface rocks. Humans are also likely to have additional radiation loads because of medical X-rays, although great care is taken to protect the gonads during these necessary irradiations. The above-ground testing of explosive nuclear devices which occurred in the 1950s added fission products such as [90]strontium and [137]caesium to the upper atmosphere from where they gradually settle, as radioactive fall-out, on the earth's surface to provide a continuing source of radioactive contamination. It has been estimated that 60 per cent of these radioactive elements will still be present at the beginning of the Twenty-first Century.

Cells vary in their sensitivity to irradiation according to which stage

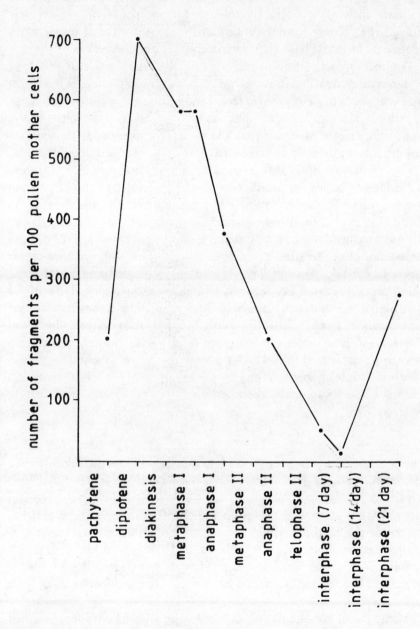

Figure 3.11 Relationship between the number of chromosomal fragments induced in meiotic cells of *Trillium erectum* at a specific dosage of radiation (50r) and the stage at which radiation occurred *Adapted from Gardner.*

of the cell cycle they have reached at the time they are irradiated. A H Sparrow, in work with the monocotyledonous plant *Trillium erectum*, showed that chromosomal abnormalities were produced much more frequently at metaphase than at interphase. See figure 3.11.

The **effect of irradiation** can also be altered by variables such as **oxygen concentration**. Low oxygen concentration decreases mutation rate, whereas a high oxygen concentration during irradiation can magnify the irradiation effect. There is also evidence that some of the mutagenic effects of irradiation are indirectly produced. Proteins or amino acids are directly affected and these then act as mutagenic agents.

The **radiation dose rate** is another important variable. Chronic, low-intensity exposure produces a smaller effect than the same total dose applied over a short period of time. Russell and his co-workers found that exposure of mouse spermatogonia to ninety roentgens per minute produced about four times the number of mutations as did the same amount of radiation given over a week. This suggests that repair processes can take place between low-intensity irradiations, but that the repair processes cannot cope, or may themselves be damaged, during or after high doses of irradiation. (Radiation applied to tissues is measured in **roentgens**. 1 roentgen – 1r – is the intensity of radiation which will give 8.77×10^{-3} joules per kilogram of air. Radiation absorbed dose used to be measured in rads, but is now measured in **grays**. 1 gray – 1 Gy – is 1 joule of absorbed energy per kilogram of absorbing material. 1 rad is 1×10^{-2} joules of absorbed energy per kilogram of absorbing material.)

Chemical mutagens

There has been much research into chemical mutagens during and since the Second World War. However, even as early as 1775 the British physician Sir Percival Pott suggested a causal relationship between an accumulation of soot in the groin region of chimney sweeps and the cancer of the scrotum which was fairly common amongst them. A similar **carcinogenic** effect was shown in 1918 by two Japanese investigators, Yamagiwa and Ichikawa. They demonstrated that skin cancers could be induced in rabbits by the repeated application of coal tar to their skin.

A long list of mutagenic substances has now been compiled. Some of them, such as nitrous acid and base analogues, have specific effects; others seem to induce a general instability of some kind which results in chemical changes within the DNA. Nitrous acid can remove an amino group from adenine to leave a different base, hypoxanthine. Adenine pairs with thymine but hypoxanthine behaves like guanine and so pairs with cytosine. Base analogues are substances which are sufficiently similar to the normal bases to be incorporated into DNA, but sufficiently different to cause base mispairing. 5-bromouracil (a pyrimidine analogue) resembles thymine and can be substituted for it (at least in bacteriophage DNA) and so what was an A-T pair becomes A-Bu. The Bu will subsequently pair with guanine and so the coding is altered.

The relationship between mutagenesis and carcinogenesis forms the basis for the **Ames test** in which potential carcinogens may be tested on

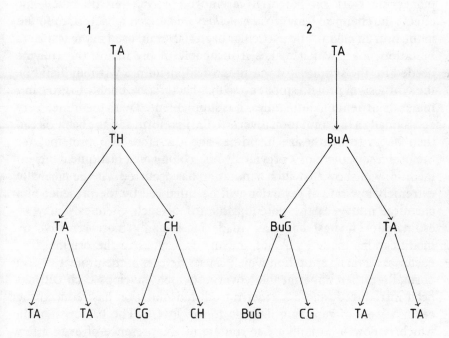

Figure 3.12   Action of two chemical mutagens

A adenine   T thymine   C cytosine   G guanine
H hypoxanthine   Bu bromouracil
1   nitrous acid deaminates adenine to form a different base, hypoxanthine. This cannot pair with thymine but can pair with cytosine. The cytosine then pairs with guanine.
2   bromouracil is an analogue of thymine – it is sufficiently like thymine to be substituted for it. However it then pairs with guanine.
In both cases the coding is altered.

a bacterial culture in a few hours or days, rather than on laboratory animals in a test which may take years. The problem of testing new substances is an urgent matter; about 500 to 1000 new chemicals are put on the market each year to join the 50 000 man-made chemicals which are currently used in commerce and industry. Although malignant tumours in animals consist of billions of cells, the original malfunction was in a single cell which subsequently grew and divided without the constraints which affected its neighbouring cells. Bacteria are single cells and so a comparison can be made between their behaviour when treated with a chemical under test, and what would happen at the single cell level in more complex organisms such as mammals.

Bruce N Ames of the University of California at Berkeley based his pioneering work on potential carcinogen testing on the mutagenic effects the chemical had on *Salmonella typhimurium*, a bacterium found in the human colon. The particular bacterial strain used in the test has a mutation, his⁻, which makes it incapable of producing the enzyme needed for the synthesis of the amino acid histidine. Without histidine the synthesis of protein stops and so his⁻ bacteria are unable to grow in a mineral nutrient medium unless it is supplemented with histidine. Very occasionally a his⁻ mutation reverts to the his⁺ form and the bacteria can then begin to synthesize histidine, and therefore also proteins, for themselves. Growth of a previously his⁻ colony on a histidine deficient medium will show that such a mutation has occurred. As the normally extremely low rate of reversion will be enhanced by the presence of a chemical mutagen, the potential hazard of such chemicals may be estimated. The test has been made increasingly more sensitive by modifications made by Ames and his colleagues to the original his⁻ bacterial strain. A mutation which causes defects in the protective and normally rather impermeable covering which envelops each cell has been introduced, and a further introduced mutation has reduced the cell's powers of repairing damage to the DNA. The bacterial strain which is now used is likely to mutate in the presence of even a few molecules of a carcinogen. However, a substance which has a carcinogenic effect in a bacterial cell is not necessarily going to produce the same effect in mammals. On the other hand, some chemicals which do not produce mutations in bacteria may be so altered by some mammalian metabolic process that they become carcinogenic. So Ames added an extract of rat liver to the test medium so that a wide range of mammalian enzymes would be present and the test would show more clearly what might happen in man.

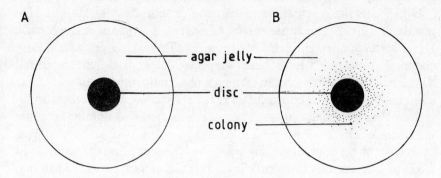

Figure 3.13 The Ames test to detect mutagenesis

A a mixture of rat liver (to supply mammalian enzymes) and the tester bacteria (his⁻) is plated on to agar jelly which does not contain histidine. The suspected carcinogen is placed on a filter paper disc on top of the plated bacteria.

B 2 days later. All the bacteria die except those in which mutations have occurred making them his⁺ and so able to synthesize their own histidine. The mutant bacteria proliferate and make visible colonies.

It is estimated that 2500 chemicals have been subjected to the Ames test. Other tests for mutagenicity using *E. coli* as the tester bacteria are also used. Sometimes the test results have demonstrated carcinogenic potential in substances which had previously been cleared in standard tests with laboratory animals. The antibacterial food preservative furyl furamide, AF2, was added to a large range of common foods in Japan after tests on rats in 1962 and mice in 1971 had shown that it was apparently safe. However, in 1973 Sugimura and his colleagues, by means of the Ames test, demonstrated the mutagenic effect of the very tiny amount of AF2 in one slice of fish sausage, and the chemical was withdrawn from the market. Ames and his colleagues were able to show that 89 per cent of the 169 hair dyes which were on sale in 1975 in the USA contained compounds which were mutagenic in bacteria. The formulae for the dyes have been modified as a result, but as tens of millions of people in the USA regularly dye their hair, the carcinogenic load in the previous years must have presented a considerable risk.

One of the great problems in identifying carcinogens which affect man is that many cancers only become evident twenty or thirty years after the exposure which initially instigated their growth, as seems to happen with the relationship between some types of lung cancer and the inhalation of asbestos fibres.

Other chemicals which may produce genetic damage do not give positive results in the Ames test because they do not affect the precise DNA sequence which codes for the histidine making enzyme. Many chemical substances have been suspected of causing chromosomal damage in man, although the evidence obtained from various tests has often turned out to be contradictory. LSD (lysergic acid diethylamide) is used therapeutically in some mental disorders and illicitly by some members of the general public. Although some studies have shown otherwise, there is an increasing amount of evidence that the addition of LSD to human cells in culture increases the amount of chromosomal damage. Evidence from individual LSD users is very variable, partly because of the lack of any standards of dose rate and dose purity, and also because many of the subjects investigated used other drugs as well. No definite conclusions may be drawn at present but the possible risk of chromosomal damage in illicit users of LSD, and of damage to their descendants, should not be ignored. The same sort of variability of test results is shown by marijuana and indeed by nicotine. Both substances have been shown to depress mitosis rates in human cells grown in culture, and human lung fibroblast cells in culture have shown various mitotic abnormalities when exposed to the two substances.

The artificial sweetener, sodium cyclamate is reported to produce chromosome breaks in cultured human leucocytes, although its presence in drinking water given to mice seemed to have no effect on the spermatocytes they produced. DDT has been shown to be associated with an increased number of chromosomal abnormalities in mice, but not so far in other mammals. Caffeine was suspected of causing chromosome damage but did not do so in healthy volunteers who ingested caffeine in four doses of 200 milligrams per day for a month. However, it has been shown to reduce crossing-over frequency in yeasts, and to increase the number of chromosome breaks in cells of the broad bean, *Vicia faba*.

Other chemicals are known which cause genetic damage in one organism but not in another. Phenols produce chromosome breaks in the cells of higher plants and point mutations in *Drosophila*, but seem to be non-mutagenic in *Neurospora*. Hydrogen peroxide is mutagenic in *Neurospora* but has no apparent effect on the cells of higher plants.

As the bases found in the DNA of all organisms are similar and differ only in their arrangement, it may be that a difference in the repair mechanism, either in form or efficiency, determines whether or not a chemical is capable of inducing an overt mutation.

Given then that an enormous range of environmental chemicals and many types of environmental radiation can induce mutation, one might expect that all genes stood an equal chance of undergoing such a chemical change. Each mutation is a chance event – it is unknown whether any particular gene will mutate at any replication. However, it appears that some genes mutate more than others.

As can be seen in figure 3.14, some genes are more likely to mutate than others. The improved methods now used in electrophoresis have suggested that there is a class of structural proteins which are almost invariable and which therefore show either low mutability of their coding genes or a rapid elimination of mutants by selection; there is a further class of proteins, soluble enzymes in particular, which show a high level of genetic variation. Structural proteins need to be highly invariant as they lock together in precise ways. Membrane bound enzymes must also be invariant as they must fit the constraints of their sites. Soluble enzymes like xanthine dehydrogenase, however, can be variable as long as any amino acid changes do not occur in the active site region, and do not change the general properties of the molecule too much. Animals which maintain large populations are therefore likely to have many variants of these alleles coding for these soluble enzymes.

Organism	Trait	Mutation per million cells or gametes
Escherichia coli (colon bacillus)	streptomycin resistance lactose fermentation	4×10^{-4} 2×10^{-1}
Neurospora crassa (red bread mould)	adenine independence	4×10^{-2}
Zea mays (maize)	shrunken seeds purple seeds	1 10
Drosophila melanogaster (fruit fly)	white eye yellow body	40 100
Mus musculus (house mouse)	brown coat piebald coat	8 30
Homo sapiens (man)	haemophilia A achondroplasia	30 40 – 80

Figure 3.14   Rates of mutation   Data from Dobzhansky, Ayala, Stebbins, Valentine.

Organism	Percentage of DNA in various frequencies		
	% unique	% copied	Number of copies
Sea urchin	38		
		25	20 – 50
		27	250
		7	6000
South African Clawed toad	54		
		6	20
		31	1600
		6	32 000
Cow	55		
		38	60 000
		2	1 000 000

Figure 3.15 Frequency of unique and repetitive DNA sequences in the genomes of three organisms *Data from Davidson and Britten.*

It has become clear that organisms vary in the actual amount of DNA they have in their cells. Mammal cells contain 1000 times more DNA than bacteria, yet there is no evidence that they produce 1000 times the number of different proteins. The basic chemistry of metabolism is similar in all organisms and so it looks, at first, as if there is just too much DNA in mammals. In prokaryotic cells each gene is present once; in eukaryotic cells many genes may be repeated a large number of times in each **genome**. Some of this repetitive DNA codes for ribosomal RNA and transfer RNA, both of which are transcribed from the code. However, unlike messenger RNA they remain in their nucleic acid form and are not used as intermediate codes to be translated into proteins. Other genes present in multiple copies code for histone proteins and for antibodies. The amount of repetitive DNA varies according to the species.

CONTROL SYSTEMS OF GENE ACTION IN PROKARYOTES AND EUKARYOTES

Research into the structure and function of the genome has made it more and more obvious that the concept of long lines of genes, arranged like beads in a necklace, each gene being 'for' a particular character is inaccurate. Genes that are transcribed are separated from each other by non-transcribed sequences. Although the control system for gene transcription has been elucidated, by Jacob and Monod, for prokaryotic cells it is still not clear how eukaryotic genes are controlled.

In prokaryotes a regulated group of adjacent genes (an **operon**) is controlled by repressor molecules which are encoded in a different gene (the **repressor gene**) which may be some distance away. The repressor molecules are affected by **metabolites** within the cell in such a way that the operator gene at one end of the operon is either switched on or off, and the rest of the operon is transcribed or not accordingly. In eukaryotic cells the system is more complicated. Bacterial mRNA molecules have a short life of only a few minutes before they are broken down and so the formation of repressor molecules and enzymes is under reasonably fine control. Eukaryotic mRNA molecules can last for hours or even days, however, and can go on signalling for protein formation for all of this time, and so the switching on or off of genes could not have as prompt an effect as in bacteria. This suggests that any control must apply at **translation** rather than at **transcription**. It has been found that mRNA may be present in cells such as the oocytes of various animals without being translated, although translation occurs if this mRNA is removed from the cells and tested *in vitro*. So in the cell either the mRNA is blocked, or the ribosomes are somehow prevented from using it.

Most eukaryotes are multicellular organisms in which cells function differently in different organs, although all the cells have received identical sets of genes by means of mitotic division during development from the original cell. There must be some mechanism for switching whole sets of genes on and off in specific cells. It has been found that functional mRNA does not always 'match up' with the DNA sequences from which it was transcribed; certain sequences of nucleotides have been removed either from within the primary mRNA or from one or other end. The DNA sequences which are going to be translated into the amino acid order have been called **exons**; those which are going to be removed by splicing in the mRNA have been called **introns**. The splicing is carried out by means of specific enzymes. It is possible that this splicing of the primary mRNA to give functional mRNA which can be translated might occur in those cells where the gene must function but not in other organs. See figure 3.16.

A slightly similar process of splicing, but in this case of DNA rather than mRNA, has been shown to be of great importance in controlling gene action in both prokaryotic and eukaryotic cells. The original

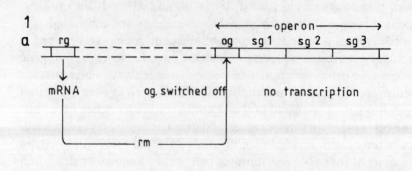

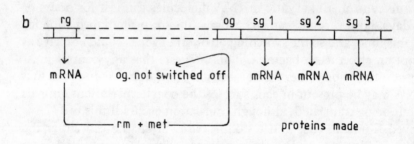

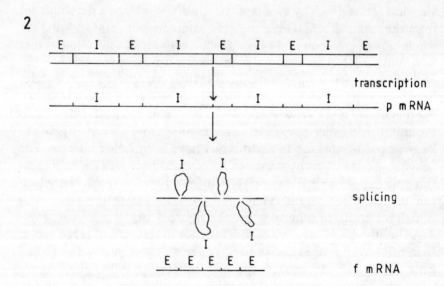

report of the process was made in the late 1940s by Barbara McClintock who had been studying the inheritance of colour and the distribution of pigmentation in maize plants. She found that particular genes were being switched on and off at abnormal times – to produce mottled kernels – a pattern of variegation which could be followed from generation to generation and which could be analysed like other heritable traits. She explained her results by suggesting that there were distinct genetic units, which she called **controlling elements**, which could apparently move from site to site in different maize chromosomes where they could sometimes act as novel biological switches simultaneously turning collections of genes on or off. However, the significance of McClintock's work was not recognized at the time, although its importance has since been acknowledged by the award of the Nobel prize in 1983.

Figure 3.16   Possible methods of gene control in prokaryotes and eukaryotes

1   Prokaryotes
     rm   repressor molecules
     rg   repressor gene
     og   operator gene
     sg 1   sg 2   sg 3   structural genes
     met   metabolites
     The diagrams show how the enzymes to deal with a particular substrate (such as lactose in the medium) are produced when needed. Repressor molecules are produced under instruction from a repressor gene. In (a) the repressor molecules combine with the operator gene and switch it off. If the operator gene is switched off, then none of the structural genes in the operon are transcribed. In (b) metabolites (such as lactose molecules) inactivate the repressor molecules by combining with them. The operator gene is not switched off; the structural genes are transcribed and the needed enzymes are made.

2   Eukaryotes (split genes)
     I   introns
     E   exons
     p mRNA   primary messenger RNA
     f mRNA   functional messenger RNA
     The DNA may contain sequences, the introns, which will not appear in the functional mRNA. Those sequences which will be translated into the amino acid order, the exons, are retained while the introns are removed by splicing from the primary mRNA.

Twenty years later other workers found a mutation in a gene in *E. coli* which seemed to have effects beyond the boundaries of the gene itself. When DNA sequences containing the mutated gene were inserted into bacteriophages and the density of the 'phage compared with that of those carrying normal genes, it became apparent that the mutated DNA sequences were heavier (and therefore longer) than normal DNA. The mutation had resulted from the insertion of a DNA fragment, which could be up to 2000 nucleotides long, and which had interrupted the normal activity of the gene and turned it off. The inserted DNA was named an **insertion sequence** or **IS element**.

Insertion sequences were then shown to be involved with the transfer of resistance to antibiotics from one bacterium to another. Bacteria contain, in addition to the DNA arranged to make a single circular chromosome, other small DNA circles called **plasmids**. It was shown that resistance to antibiotics such as penicillin and ampicillin was because of a plasmid gene coding for a protein which inactivates the antibiotic. The transfer of resistance from one plasmid to another was always accompanied by an increase in size of the plasmid and in 1974 Hedges and Jacob suggested that the gene for ampicillin resistance was carried by a DNA element which could be 'transposed' from one plasmid to another and which they called a **'transposon'**. The transposon containing the ampicillin resistance gene was found to be a 4800 nucleotide segment which moved as a discrete structural unit. The two ends of this segment had a unique structural feature – they were found to have nucleotide sequences which were complementary to each other but in reverse order. If the two strands of the plasmid DNA were separated in the laboratory and then allowed to re-anneal, the transposon would make a stem-and-loop structure which could detach itself and move into another plasmid.

Resistance to many different antibiotics could therefore move from plasmid to plasmid. The resistance gene could also move from bacterium to bacterium by the processes of transformation (whereby 'foreign' DNA may be picked up from dead bacteria in the surrounding medium) and transduction (whereby the 'foreign' DNA is introduced by bacteriophages), and this fact has produced many medical problems since the introduction of antibiotics in the 1940s.

Numerous examples of transposable elements have now been found in higher organisms such as *Drosophila* and the yeast *Saccharomyces cerevisiae*, and work is continuing in many laboratories around the world. Recent work by Tonegawa at the Basel Institute for Immunology suggests that the ability of mouse cells to produce specific antibody molecules in response to the injection of foreign protein (antigen) depends on chromosomal rearrangements.

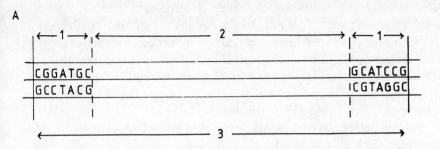

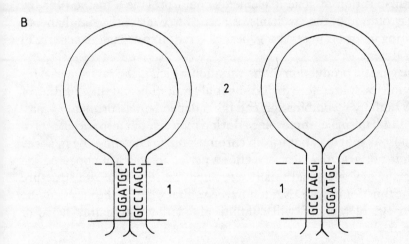

Figure 3.17   Stem-and-loop structures formed from transposable elements in plasmid DNA

1   inverted repeat sequences of nucleotides at the ends of the transposable element
2   internal genes of transposable element
3   transposable element
A   Section of plasmid DNA
B   If the two strands in the DNA separate, the two ends of each strand of transposable element can then pair up making a stem-and-loop structure. This can then detach itself and move into another plasmid.

Although transposition in eukaryotic cells is not brought about by precisely the same methods as in prokaryotes, it achieves the same result – sections of DNA which normally have no connection with each other are brought together. Gene shuffling in the normal process of crossing-over during meiosis results in the exchange of equivalent sections of homologous chromosomes. Transposable elements can join DNA sections with little ancestral relationship and can, if they move into suitable positions, have great effects by acting as switches for their new neighbours. Their movement is not completely random; there are certain sites in the DNA ('hot spots') which are more likely to receive transposons. Transposons also contain genes which code for the enzymes which bring about their transposition.

Such genetic rearrangements can be of great biological importance on **two time scales**. On a **developmental** scale the results are shown within a single generation, whereas on an **evolutionary scale** the effects may only be seen after several generations. McClintock has suggested that the transposition of genetic elements could provide a method for the rapid evolution of the mechanisms needed to control the simultaneous switching on or off of several genes, as is required during the course of development.

The genetic production of the variation which is the raw material for evolution must therefore include not only mutation and the shuffling of alleles both by recombination and the independent assortment of maternal and paternal chromosomes during meiosis, but also the effects of transposons moving about from chromosome to chromosome in a way which is not yet fully clear. So after a period in which it appeared the molecular biologists had finally extracted the secrets of the structure and the functioning of DNA it now seems that what has been called 'the golden age of biology' will continue into other fascinating fields of research.

SUMMARY

Mathematical inspiration, supported by long years of careful experimentation, allowed Mendel in his work on heredity in pea plants in the middle of the Nineteenth Century to lay the foundations of the science of genetics, a science which has flourished throughout this century. More and more has been learned of the way in which cells, and organisms, depend for their reproduction on the properties of the nucleic acids. These self-replicating molecules, DNA (deoxy-ribonucleic acid) and RNA (ribonucleic acid), contain all the information, in the coded pattern of their organic bases, needed for the machinery of the cell to turn out the proteins which are needed, as enzymes, to enable all the chemical reactions of the cell to proceed. The genetic code determines the arrangement of amino acids in the proteins, and therefore determines the properties of the proteins. Mutations are changes in the code which may alter its instructions and cause different proteins to be produced. Detection of these different proteins could therefore show the occurrence of mutation. The most useful way of searching for mutation by means of changes in proteins is by electrophoresis, a process which relies on the behaviour of charged proteins in an electric field.

It has now become obvious that the original idea of long strings of separate genes, each affecting one specific character, is incorrect. Often many genes are involved in one character; sometimes many characters are affected by a single gene. Some genes mutate to give many possible alleles, any two of which may be present in a normal diploid cell. The rate at which genes mutate is variable, some genes mutating more than others. The rate is also affected by such environmental hazards as radiation or particular chemicals. In some organisms each gene is present once; in others multiple copies of genes are present. The fine details of the ways in which genes can control the metabolism of the cell are not yet clear. Various types of genetic switches such as the repressor/operator systems in prokaryotes, or split genes and transposons in eukaryotes have been suggested.

FURTHER READING

Ames, Bruce, Kammen, H O, Yamasaki, Edith, 'Hair dyes are mutagenic: identification of a variety of mutagenic ingredients', *Proc. Nat. Acad. Sci. USA*, vol. 72 (1975) pp. 2423–2427.

Ayala, Francisco J, 'The mechanisms of evolution', *Scientific American* (September 1978).

Chambon, Pierre, 'Split genes', *Scientific American* (May 1981).

Cohen, Stanley N, Shapiro, James A, 'Transposable genetic elements', *Scientific American* (February 1980).

Davidson, E H, Britten, R J, 'Organisation, transcription and regulation in the animal genome', *Quart. Rev. Biol.*, vol. 48 (1973) pp. 565–607.

de Serres, F J, 'AF-2 – Food preservative or genetic hazard?', *Mutation Res.*, vol. 26 (1974) pp. 1–2.

Devoret, Raymond, 'Bacterial tests for potential carcinogens', *Scientific American* (August 1979).

Doolittle, W Ford, Sapienza, Carmen, 'Selfish genes, the phenotype paradigm and genome evolution', *Nature*, vol. 284, no. 5757 (17 April 1980) p. 601.

Hozumi, N, Tonegawa, S, 'Evidence for somatic rearrangement of immuno-globulin genes coding for variable and constant regions', *Proc. Nat. Acad. Sci. USA*, vol. 73 (1976) pp. 3628–3632.

Hubby, J L, Lewontin, R C, 'A molecular approach to the study of genic heterozygosity in natural populations. 1. The number of alleles at different loci in *Drosophila pseudoobscura*', *Genetics*, vol. 54 (1966) pp. 577–594.

Jones, J S, 'How much genetic variation?', *Nature*, vol. 285, no. 5786 (6 November 1980) pp. 10–11.

Lehmann, H, Carrell, R W, 'Variations in the structure of human haemo-globin', *British Medical Bulletin*, vol. 25 (1969) pp. 14–23.

McClintock, Barbara, 'The origin and behaviour of mutable loci in maize', *Proc. Nat. Acad. Sci. USA*, vol. 36 (1950) pp. 344–355.

McClintock, Barbara, 'The control of gene action in maize', *Brookhaven Symposia in Biology*, vol. 18 (1965) pp. 162–184.

Rosamond, John, 'Special sites in genetic recombination', *Nature*, vol. 286, no. 5770 (17 July 1980) pp. 202–203.

Singh, R S, Lewontin, R C, Felton, A A, 'Genetic heterogeneity within electrophoretic "alleles" of xanthine dehydrogenase in *Drosophila pseudoobscura*', *Genetics*, vol. 84 (1976) p. 609.

Stern, C., *Principles of Human Genetics* (3rd ed.) (W.H. Freeman and Co., San Francisco, 1973).

Yahagi, T, Nagas, M, Hara, K, Matsushina, T, Sugimura, T and Bryan G T, 'Relationships between the carcinogenic and mutagenic or DNA-modifying effects of nitrofuran derivatives, including 2-(-furyl)-3-(5-nitro-2-furyl)acrylamide, a food additive', *Cancer Research*, vol. 34 (1974) pp. 2266–2273.

Yanovsky, Charles, 'Gene structure and protein structure', *Scientific American* (May 1967).

4

The genetic structure of populations

In chapter 3 we have been considering ways in which the genetic material of individual organisms may be altered by changes in composition, arrangement or quality. Such changes occur in a genome which the individual acquired from reproduction in the parental generation. Each organism is descended from one or two individuals of the previous generation. If it has been asexually reproduced by the branching away of a part of the parent which has enlarged by means of mitoses, then the offspring will at least start its life genetically identical to the parent. So the parent, and its offspring, and their offspring and so on will form a collection of organisms with the same genome and which form a clone. Of course, mutation may occur at any time in any of these organisms, and some mutations may be carried down in the same asexual way to produce different clones.

If the organism has been sexually reproduced from two parents it is the result of the chance shuffling of the genes of its parents during the formation of their gametes, and the chance union of two gametes to make the zygote from which it developed. It is like a game of cards played on an enormous scale. In card games a pack of fifty-two cards, made up of four suits each of thirteen cards, is shuffled (with different techniques and with different degrees of skill and artistry depending on the 'class' of the players!) between each game. A random hand is then dealt to each player and the particular game may proceed. At the end of each game, the cards are again put into the 'pool' of cards ready to be reshuffled. Each sexually reproduced organism is 'stuck' with the particular random combination of genes dealt out to it, just as if it were a card game, at the moment of fertilization. However, when it, in its turn, reproduces it will pass a random mix of its genes, in its gametes, into the pool from which the new organism will be derived. Just as the hands of cards from each player go into the card pool, the possible gametes from all the individuals with which any individual organism could mate go together into the **gene pool**.

So we are concerned, at least in sexually reproducing organisms, not just with individuals but with collections of individuals which can actually or theoretically mate with each other. Collections of this type are called **Mendelian populations**, and it is the population rather than its individual members which can persist through time. The study of the genes in such populations is called **population genetics**. It tries to describe, in mathematical terms, the consequences of Mendelian inheritance at a population rather than at an individual level.

CHARACTERISTICS OF POPULATIONS

Populations need to have certain characteristics if they are going to manage to survive for an appreciable length of time.

1　There must be a capacity for increase in number which can be related to a variable death rate to keep the population constant.
2　There must be a store of genetic variability so that the population has the capability to adapt to suit an environment which is likely to change both in time and in space.
3　There must be limits to the amount of variation in each population.

Let us now consider each of these population features.

The capacity for increase to offset a variable death rate

One of the features of many organisms which impressed both Darwin and Wallace was their enormous reproductive capacity. Gametes may be liberated in millions; zygotes may be formed in vast numbers; asexually reproduced bacteria may themselves reproduce about twenty minutes after their own formation. Biological reproduction means a geometric increase in numbers. It is obvious that not all the progeny can survive, given the resources available in the particular environment inhabited by the organisms. One of the earliest mathematical statements of the geometric increase in theoretical populations without external constraints was made by Linnaeus, who showed that if an annual plant produced two seeds, and each of these offspring produced two seeds and so on, the population of plants after twenty years would be 1 million. A geometric progression which had such a small original figure has always amazed many people. There is a story in the Arabian Nights about a man who asked for what seemed to be a very meagre reward for fulfilling some task. All he asked was one grain of wheat for the first square of the chessboard, two for the second square, four for the third, eight for the fourth, and so on to the sixty-fourth. The outcome of the progression was not reached. By the time they would have got to the last square of the chessboard, the man would have accumulated about 10^{19} grains, and if each grain had weighed 0.1 grams his reward would be about 10^{12} tonnes.

Of course, geometric population growth cannot go on for long periods of time in natural conditions. A small sample of bacteria, given the good environment provided either on nutrient agar jelly or in a liquid nutrient medium kept at a suitable temperature, will enter a geometric reproductive phase and their numbers will increase accordingly. The nutrient concentration in the medium will, however,

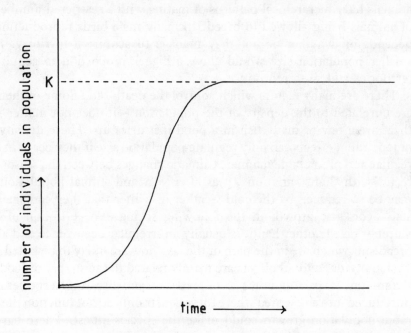

Figure 4.1 Typical population growth curve in an unexploited environment

The population is small to begin with, but rapidly increases in size. Eventually it reaches the maximum population size for the environment and becomes stable. K represents the carrying capacity.

decrease proportionately and this, and the possible accumulation of toxic waste products, will limit further population growth. Eventually a maximum population size may be reached, depending on the environmental conditions.

Realistic population projections must therefore include some figure which represents an upper limit, which is imposed by restricted environmental resources and competition with other organisms. The upper limit is known as the **carrying capacity** of the environment, and is given the symbol K. As the population nears the carrying capacity for its environment, its rate of increase is reduced until finally it stabilizes with the birth rate equalling the death rate at any specific time. Such a steady state may be maintained by negative feedback, which can control the birth rate or the death rate. Several kinds of negative feedback are known to control reproduction rates. Female rats and mice respond to overcrowding by developing abnormal reproductive behaviour in which ovulation may be suppressed, or maternal behaviour may be so inefficiently carried out that the offspring do not survive. Many social animals have hierarchical patterns of mating with a restricted number of animals being allowed to breed. In many male birds reproduction depends on whether or not they manage to acquire a **territory**. A smaller population size would allow a bigger proportion to acquire territories and so to reproduce.

There are many factors which control the death rate. Some of them are unrelated to the density of the population – it does not matter if there are a few or many fish in a pond if it dries up. These **density-independent** effects can only regulate population size if they occur in a regular and predictable manner. Climatic changes between the seasons recur, with slight variations, year after year and animal populations may be so reduced by the cold weather in winter that they can again start geometric growth in the following spring. The eruption of a volcano, on the other hand, is usually an irregular occurrence and all organisms which are in the path of the lava flow are likely to be killed.

Density-dependent effects are mainly related to the supply of food, the amount of space and the risk of predation and disease. If a species is introduced into a new area where these constraints do not function then rapid population growth could make the species a pest. There have been many examples of the accidental or deliberate introduction of species which have subsequently 'got out of hand', and which have needed enormous amounts of research and a large financial commitment before they could be controlled. The rabbits of Australia form the most quoted example of a species increasing its numbers with terrifying speed in the absence of native predators, but there are many others. Mink and coypu were brought to Britain to be reared in 'fur farms' so that their pelts could be used to make fur coats. However, there were enough escapes to allow both of these species to become established in the wild, where they have now become a considerable nuisance. The North American grey squirrel was brought to Woburn in Bedfordshire in 1890 and has now taken over most of the areas once occupied by our native red squirrel. The European starling, *Sturnus vulgaris*, was taken to the USA in 1890 by someone who thought it was an attractive bird,

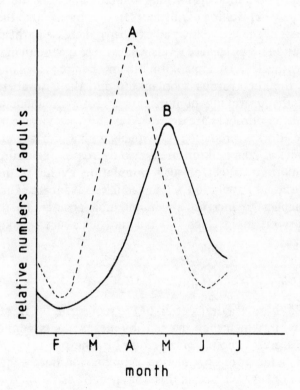

Figure 4.2 Typical cycle of phytoplankton and zooplankton in a marine ecosystem

A   phytoplankton (producers)
B   zooplankton (consumers)

The zooplankton peak in abundance about a month later than the phytoplankton they are consuming.

and introduced into Central Park in New York. Since then it has spread until it has reached the west coast. As in Britain various methods have been tried to get rid of the large starling flocks which deface city buildings and kill trees with their droppings. The mongoose which was introduced into Jamaica from India to control rats ended up eating many other things like chickens, eggs, kittens and puppies.

When predators are present and there is also a regularly fluctuating climate, populations of both the prey and the predators tend to follow cyclical patterns. The prey population begins a rapid increase because of some temporary change in the environment. The increased number of prey animals makes it easier for the predators to catch them and so the predator population also begins to increase. In marine ecosystems, for example, an upwelling of nutrients from the sea bed to the surface at certain times of the year allows an explosive increase in the numbers of phytoplankton in the surface levels where there is enough light for them to photosynthesize. This 'plankton bloom' then provides food in abundance for the herbivorous zooplankton and their numbers increase. However, the upwelling of nutrients is not permanent and so the phytoplankton numbers are decreased both by a shortage of nutrients and by the effect of predation. Obviously with a shortage of *their* food, the numbers of zooplankton must also decrease. See figure 4.2.

The important aspect of such **population cycles** is that there is always a residual population which can take advantage of any environmental changes to increase the population again. The genetic and evolutionary effects of such 'booms and busts' are considered again later.

A store of genetic variability

On page 85 there is a description of the pioneering work of Lewontin and Hubby, in which they showed, by means of electrophoresis, that there was a great amount of variability in the proteins found in the individual members of a population of *Drosophila*. Before the technique of electrophoresis was used in this way, genetic variation could only be deduced from its effect on the phenotype of the organism. Recessive alleles could be present in a population but until they came together in homozygotes their presence could not be known. Phenotypes do not 'emerge' from genotypes in a fixed, unchanging pattern, however. As you have seen in the previous chapter, it is clear that genes themselves cannot be thought of as separate entities – they need other genes which

Figure 4.3 Himalayan rabbits

A   this rabbit has been reared at a high temperature (over 25°C) and the gene
      for dark pigmentation does not function.
B   this rabbit has been reared at a lower temperature. The gene functions in
      the colder extremities such as ears, feet and tail.

control their functioning. Even when enzymes have been made according to the coded instructions of the genes, the enzymes have to function
in conditions which are basically determined by the environment in
which the organism lives.

Much of the variation which is seen both within and between populations is the direct result of different environments. An obvious effect of
a shortage of food, for example, is to produce smaller stature and lower
weight in animals which are genetically 'meant' to be more robust.
Similarly, genetically tall plants growing in soil which is deficient in
soluble nitrogen compounds will be stunted. Sometimes food plays a
different sort of phenotypic role. Certain breeds of rabbits may have
white or yellow fat. The difference is partly genetic – there are white
and yellow alleles; the yellow allele cannot function, however, unless
the pigment xanthophyll is present in the rabbit's food. So white fat in a
rabbit might be either genetically determined or the result of a 'faulty'
environment. The gene for dark pigmentation in Himalayan rabbits or
Siamese cats can function only in temperatures which are lower than
those maintained in most of the body of the animal. So the extremities
like ears, nose, feet and the tail become dark coloured while the rest of
the animal remains white. If the hair is shaved from another part of the

animal's body, then the new hair will also be dark, that region now being appreciably colder than the rest of the skin because of the loss of its insulation.

Such obvious environmental effects may be used to derive some measure of the **heritability** of a character. The **coefficient of heritability** (which is given the symbol h^2 which means the coefficient and not its square) is the proportion of the total variance of the phenotype in a particular environment which can be attributed to the genes. In other words, it tells us how the environment can alter a particular character. A heritability of 1 means that the character is purely genetically determined and that the environment has no effect, whereas a heritability of 0 would mean that all the variation in the particular character in the population is caused by the environment. As you might realize, most characters lie somewhere between these two extremes. Heritability estimates have been very useful in agriculture and animal husbandry.

Animal and trait	per cent heritability
Cattle	
Birth weight (Angus)	49
Gestation length (Angus)	35
Milk yield (Ayrshire)	43
Chickens	
Body weight (Plymouth Rock, 8 weeks)	31
Egg production (White Leghorn)	21
Egg weight (White Leghorn) .	60
Mice	
Tail length (6 weeks)	60
Body weight (6 weeks)	35
Litter size	15
Drosophila melanogaster	
Abdominal bristle number	52
Thorax length	47
Wing length	45
Egg production	18

Figure 4.4 Table to show estimates of heritability for various characters in different varieties of farm and laboratory animals *Taken from Goodenough*.

However, it must be remembered that every organism lives in a unique situation. No other organism can have precisely the same set of genetic and environmental factors affecting it in that particular place and time; nor could it have had identical environments during its development. This means that in practice it is almost impossible to create the uniform environments one would need to deduce absolutely rigid figures for heritability. A population of insects such as *Drosophila* being reared in a culture bottle on an artificial diet in fixed conditions of temperature and light might seem to be living in standardized environmental conditions but the standardization can only apply to obvious and major sources of environmental variation. There will be all sorts of environmental differences which cannot be controlled, such as the amount of food available to early or late hatching eggs, or changes in humidity as more eggs hatch, or the accumulation of toxic products as the larval population increases. All of these can have effects on the developing insects.

The environment therefore cannot be uniform, but consists of a range of features which may vary over small distances, and over short periods of time. In some cases it appears that genotypes can choose the best environment. Kettlewell found that if he released melanic and non-melanic Peppered Moths into a barrel which was internally painted with black and white stripes the moths settled on the background against which they were best camouflaged. Similarly *Anolis* lizards in the Caribbean choose backgrounds which suit their own patterning.

Sometimes the choice of habitat does not seem to be related to camouflage. Mosquitoes of the species *Anopheles arabiensis* in Nigeria which have certain **chromosome inversions** are found indoors more frequently than those from the same population which do not have these inversions. This has the added effect that indoor mosquitoes are more likely than outdoor mosquitoes to bite people, and so their choice of habitat could also alter their dietary habits. Sometimes there is a difference in food choice between different genotypes without a change in habitat being necessary. The fresh-water isopod, *Asellus aquaticus*, is polymorphic at a locus controlling amylase production. Homozygotes for one of the alleles select willow leaves on which to feed when given a choice between willow and beech leaves in a laboratory experiment much more than do the other genotypes.

So genetic variability within a population, apart from maintaining a large gene pool from which future generations may be derived, enables the population to make the best use of a very variable environment.

Limits to the amount of variation

Accepting that a store of genetic variability is very necessary for the maintenance of a population in a non-uniform environment, it is nevertheless important that this variation is not limitless. To have a very poor hand dealt to one from a card pool in a game of cards is unfortunate; the 'dealing' from a gene pool of a very poor genetic mix which is entirely ill-adapted to the local environment is wasteful of biological resources. So generally the gene pool for each population is limited in such a way as to produce reasonably well adapted organisms. Most members of a population tend to breed with others which live very close to them, which is what we ourselves have done until recently. Many people married within their own street, or village, or limited area of a town before transport became more freely available. Such small local breeding units of larger populations are called **demes**. Demes are partially isolated from other demes in the population but interbreeding occurs from time to time giving an influx of new genes.

A Mendelian population can therefore differ according to the organism. Sometimes it has wide geographic boundaries, as in the case of plants which can be wind or insect pollinated and in which the two parents may be far apart. Plants which are self-pollinated may produce local populations with limited gene pools, particularly if the seeds are not equipped for long distance dispersal. Often a population, although potentially interbreeding throughout, is composed of smaller units, the demes, within which most of the breeding actually occurs.

GENE FREQUENCIES IN AN IDEALIZED POPULATION

The Hardy-Weinberg Law

As you see, there are many variables which must be considered in the study of population genetics. In 1908, when the study of genetics itself was in its infancy, an English mathematician, **G H Hardy**, and a German physician, **W Weinberg**, independently produced a simple mathematical formula to describe gene frequencies within an idealized population. An idealized population differs from a real population in two main ways – firstly, individuals are assumed to mate completely randomly, and secondly, the population is assumed to be genetically static, with no 'new' alleles entering it from any source. The formula, now known as the **Hardy-Weinberg Law**, deals with individual genes found within such a population, whether there are two or a series of alleles at the gene locus. It specifies that if matings between members of a population are completely at random, that is if every male gamete in the pool has an equal chance of uniting with any female gamete, then the frequency of the zygotes which can arise can be predicted from the parental frequency of alleles. If p is the frequency of allele A, and q is the frequency of allele a in a population, the following could occur during random mating:

	male gametes	
	p(A)	q(a)
female gametes p(A)	p^2(AA)	pq(Aa)
q(a)	pq(Aa)	q^2(aa)

$$(p + q)^2 = p^2 + 2pq + q^2$$

The probability that the allele A will be 'drawn' from the gene pool to join with another allele to make a zygote is indicated by p, and similarly the probability that the allele a will be 'drawn' is indicated by q. To estimate the probability of a particular zygote mix we just multiply the probability for each particular allele. For AA this would be p $\times$ p (p^2); for Aa it would be p $\times$ q + p $\times$ q (2pq); for aa it would be q $\times$ q (q^2). Suppose 20 per cent of the alleles are A and 80 per cent are a. Then:

4 per cent of the zygotes are AA $\left(p^2 = \dfrac{20}{100} \times \dfrac{20}{100} = \dfrac{4}{100} \right)$,

64 per cent are aa $\left(q^2 = \dfrac{80}{100} \times \dfrac{80}{100} = \dfrac{64}{100} \right)$,

and the rest are Aa.

In some cases co-dominance between alleles occurs, and then the proportions of the three genotypes will be readily obvious, each genotype producing a different phenotype. However, when dominance means that the dominant homozygote and the heterozygote share the same phenotype, the Hardy-Weinberg Law can allow the proportions of heterozygotes in the population to be calculated. If, for example, we

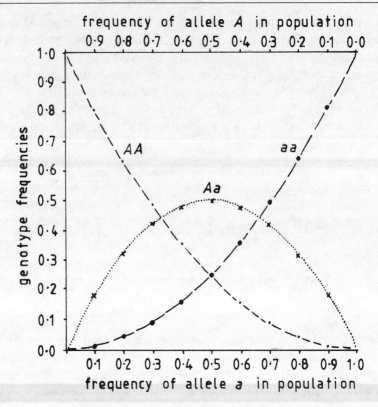

Figure 4.5   The effect of changes in gene frequencies on frequencies of genotypes in a population

A decrease in the frequency of allele A in the population means an increase in the frequency of allele a (assuming that these are the only alleles involved). A decrease in the frequency of an allele produces a decrease in the frequency of its homozygote genotype. So as allele A decreases in frequency from 1 to 0, the frequency of genotype AA also decreases. The related increase in the frequency of allele a means an increase in the frequency of genotype aa. Genotype Aa reaches its highest frequency when the alleles A and a are equally frequent in the population.

find that 36 per cent of people in a randomly breeding human population cannot taste the substance phenylthiocarbamide (PTC) which the rest of the population finds very bitter to taste, and we also know that the phenotype for tasting or otherwise is controlled by a single gene locus, and that the taster allele is dominant to the non-taster allele, then we can conclude that:

1 the frequency of the non-taster homozygotes (tt) which is 36 per cent corresponds to q^2, and so q must be $\sqrt{0.36} = 0.6$;
2 the value of p must therefore be $1 - 0.6 = 0.4$;
3 the remaining 64 per cent of the population who are tasters is made up of 16 per cent dominant (TT) individuals (p^2 when p = 0.4) and 48 per cent heterozygotes (Tt) (2pq where p = 0.4 and q = 0.6).

The Hardy-Weinberg Law may be used to calculate the percentage of members of a population who are heterozygous for a recessive deleterious allele. The allele for human albinism, for example, is recessive and the condition appears in the proportion of 1 albino to 20 000 people with normal colouration. This proportion of 1 in 20 000 is represented by q^2 and so q is

$$\sqrt{\frac{1}{20\ 000}} = \frac{1}{141} \text{ (approximately).}$$

p must therefore be

$$1 - \frac{1}{141} = \frac{140}{141}$$

and

$$2pq = 2 \times \frac{140}{141} \times \frac{1}{141} = \frac{280}{20\ 000} \text{ (approximately).}$$

So there are only 280 people in a population of 20 000 who are heterozygous, and the chance of them mating is very low. Even if two heterozygotes do mate, there is only a one in four chance that any of their offspring will have the homozygous recessive genotype which would produce albinism. People can be reassured by the low probability that they may be carriers of the recessive allele.

The probability of being a carrier of the recessive allele which causes sickle cell anaemia when homozygous is relatively high in the black population of the USA. Here one person in 400 is homozygous for the allele; by using the Hardy-Weinberg calculation it can be shown that about one in ten of the population are heterozygous. In many examples of genetic disorder, natural selection has tended to eliminate recessive homozygotes, or to prevent them from reproducing. Some geneticists in the late Nineteenth and early Twentieth Centuries thought that it would be reasonable and wise to give natural selection a 'helping hand' in trying to improve the human race by means of a programme of eugenics (the term 'eugenics' having been invented by Francis Galton in 1883). After all, man had been improving his domesticated animals and plants over many centuries by means of deliberate selection. Why not extend such a plan to his own species? Two policies were advocated. **Positive eugenics** involved the selective breeding of the 'best' type of human beings to produce an elite race. **Negative eugenics** meant checking the birth rate of the 'unfit' so that they would produce very few, or no, offspring to carry their 'bad' genes into the next generation. You should be able to see the flaws in the arguments. Who is going to decide what is the 'best' type of human being? What does 'best' mean, anyway? An adaptation which might be useful in particular circumstances might well lose its usefulness if the environment alters. The whole idea of deciding to breed a 'master race' is morally indefensible, and it is an aim which would be impossible to achieve in practice. Negative eugenics, if this involves the compulsory sterilization of those considered 'unfit' is also morally indefensible. A consideration of the examples given of the working of the Hardy-Weinberg Law should also persuade you that a programme to eliminate deleterious recessive alleles from a population by means of the elimination or sterilization of recessive homozygotes is completely inefficient. The presence of heterozygotes in a population (whether in a relatively high proportion such as the 10 per cent for the sickle cell allele in the black population of the USA, or the much lower proportion of 1.4 per cent for the allele for albinism) means that new homozygotes must constantly emerge. In many cases the affected homozygotes would be unlikely to produce offspring anyway, and so their compulsory sterilization would have no effect on future generations.

Although we have used examples of three human populations to show how the Hardy-Weinberg Law may be applied, it is obvious that these are not idealized, but real, populations and therefore do not fulfil the requirements of the Hardy-Weinberg equilibrium. Nevertheless the calculation can yield useful information. (Some worked examples of Hardy-Weinberg calculations are given in Appendix A.) The Law makes a good starting point and helps to define some of the important genetic variables in a population. The Hardy-Weinberg Law enables us to calculate **population statics**; **population dynamics** are needed to cope with the world of **real populations**. R A Fisher, Sewall Wright and J B S Haldane were amongst the first to suggest the more complicated mathematical models needed for the fluctuations of allele frequencies shown in real populations, and biomathematicians have continued to produce models which have become more and more complicated as new data from natural populations have emerged.

Conditions needed for the Hardy-Weinberg Law to apply

As we have said, the Hardy-Weinberg Law describes static allelic frequencies in a population – the population is not changing from generation to generation and so is *not* evolving. There are certain necessary conditions for such an equilibrium to be established and maintained.

1   **Meiosis must be normal** so that chance is the only factor in gametogenesis. Heterozygotes, Aa, must produce equal proportions of gametes containing A or a.
2   There must be no **introduction of new alleles by mutation**, or mutation must be balanced – each time allele A changes into allele a somewhere in the population, allele a changes into allele A somewhere else.
3   The population is **closed** with **no immigration or emigration**.
4   The population must be **infinitely large**, and all **mating is at random**.
5   **No selection** is operating – there is neither differential mortality nor differential reproduction.

If any of these conditions is not fulfilled, the population is likely to move away from equilibrium and the gene frequencies will then alter in subsequent generations. Some definitions of evolution report only the fact of such a change in gene frequencies. Sewall Wright in 1942 stated that 'evolution is the statistical transformation of populations'. Many more recent definitions have been given, most of which include the concept of adaptation to a changing environment. Richard Lewontin has described evolution as a process which converts variation *within* a population into variations *between* populations both in space (the formation of races and species) and in time (the evolution of phyla).

We must therefore consider in more detail how the five conditions for the Hardy-Weinberg equilibrium may be disturbed and so bring about evolution in the population, and this we shall do in the next chapter.

SUMMARY

Although the individual organisms which make up a population are mortal, the population itself can keep going as new organisms are born to replace those which have died. It is obvious that as the **death rate** may be very variable, depending as it does on **environmental factors** such as changed temperature or a shortage of food, there must be a **capacity for increase** to keep the population size constant. The population will survive if there is **sufficient variation** in its members to ensure that they are not all killed because of a change in the environment whether this occurs in space or in time. So a store of genetic variability is needed within each generation (to offset environmental change in space) and between one generation and the next (to offset environmental change in time).

The genetic variability may be measured in terms of **gene frequency** within the population. The gene frequency gives the proportions of the different alleles of a specific gene in all the organisms in the population. The study of gene frequencies and the way in which they may change is **population genetics**. In **natural populations** the gene frequency is likely to change from generation to generation as a result of many different forces. However, much useful information about gene frequencies and their effect on the numbers of homozygotes and heterozygotes may be obtained by considering an **idealized population** in which gene frequency change does not occur. For this the **Hardy-Weinberg Law** is used.

FURTHER READING

Christensen, B, 'Habitat preference among amylase genotypes in *Asellus aquaticus* (Isopoda, Crustacea)', *Hereditas*, vol. 87 (1977) p. 21.

Coluzzi, M, Sabatini, A, Petrarca, V, Dideco, M A, 'Behavioural divergences between mosquitoes with different inversion karyotypes in polymorphic populations of the *Anopheles gambiae* complex', *Nature*, vol. 266, no. 832 (1977).

Jones, J S, 'Can genes choose habitats?', *Nature*, vol. 286, no. 5775 (21 August 1980) p. 757.

Penrose, L S, 'Ethics and eugenics', in Fuller, Watson, (ed.), *The Social Impact of Modern Biology* (Routledge and Kegan Paul, 1971).

Schoener, T W, Schoener, A, 'The ecological context of female pattern polymorphism in the lizard *Anolis sagrei*', *Evolution*, vol. 30 (1976) p. 650.

5

The mechanism of evolution

Evolution being a process in which the gene frequencies of a population change from generation to generation, it is obvious that it is not shown by an idealized population to which the Hardy-Weinberg Law applies. As we have listed at the end of chapter 4, there are various ways in which gene frequencies change in real populations.

CHANGES IN MEIOSIS

Normally in meiosis heterozygotes (Aa) will produce equal proportions of gametes containing the allele A and those containing the alternative allele a, and the gametes themselves will have equal chances of joining in fertilization and so subsequently developing. However, it now appears that the normal segregation ratios may be disturbed. There may be abnormalities in meiosis, collectively named **meiotic drive**, which make it much more likely that a gamete will contain one rather than the other allele. The gametes, particularly the male gametes, which have been made may also differ in their longevity or their activity and so some have a higher chance of reaching and fertilizing the female gamete. It is also possible that some zygotes can survive better than others through their developmental stages to become viable offspring.

Consider the *tailless* locus T which is on chromosome *XVII* in the house mouse, *Mus musculus*, and which has been extensively studied by Dunn and Bennett, and more recently by Arntz and Jacob. The gene has a series of multiple alleles. The wild type allele, $+$, may be induced to mutate into the dominant form T in the laboratory. Mice which are heterozygous, $+T$, have short tails; homozygotes, TT, die in the uterus during the gestation period. There are also other alleles at this locus, t, which are fully recessive to the wild type. They do, however, have an effect when they are with the allele T. Tt heterozygotes are tailless. There are many different t alleles and some, when homozygous, prevent development for various reasons — t^{12} homozygotes, for example, fail to form blastocysts and t^{W37} homozygotes fail to implant in the uterus. Such lethality in t homozygotes would quickly lead to the elimination of t alleles from the population but for the fact that males with the genotype $+t$ produce 95 per cent spermatozoa containing the allele t, and only 5 per cent spermatozoa containing the $+$ allele. Spermatozoa bearing t alleles are longer lived than others and this raises the probability that they will fertilize an egg. t alleles are widely carried by natural mouse populations and there is some evidence that this preferential transmission of the t alleles can offset the lethality of being homozygous. In 1964 six male mice carrying the t allele were released onto Great Gull Island, a 7.3 hectare island in Long Island Sound. The t allele was not present in the resident mouse population and it has now been shown to have spread through the population. So in the case of the mouse T locus there is evidence of meiotic drive, of changes in gamete longevity, and also of differential survival of the zygotes which may be formed.

A similar alteration in the proportions of the different gametes which are made (the **segregation ratio**) results from the *segregation distorter* locus (SD) on chromosome II in *Drosophila melanogaster*. Just as with the T locus in mice, segregation is affected only in males which are heterozygous for the wild type and the distorter alleles ($+SD$), resulting in an excess of gametes bearing the *distorter* allele. The *distorter* allele seems to interact with the wild type allele during the formation of sperms (spermatogenesis) and few sperms bearing the wild type allele are formed.

Both the T locus and the SD locus are in autosomes (non-sex chromosomes) and, although they may affect gene frequencies in the next generation, they do not affect sex ratios. However, sex-linked distorter genes are known to occur in some organisms. In the mosquito *Aedes aegypti* a gene, M^D, on the Y chromosome can induce breakage of the X chromosome during meiosis. Spermatids bearing the faulty X chromosome degenerate, leaving only the Y bearing spermatids to develop into spermatozoa. This, of course, would lead to populations made up entirely of males and to subsequent extinction, except that certain X chromosomes seem to be able to resist the M^D induced breakage. So far eight categories of X chromosomes, ranging from highly sensitive to highly resistant to breakage, have been identified. Sex ratio alteration is also found in certain races of *Drosophila pseudoobscura* where X and Y chromosomes do not always pair in meiosis. When this happens the Y chromosome degenerates; the X chromosome undergoes a further replication and so all the spermatozoa will carry an X chromosome and only daughters will be produced.

So meiotic drive and other factors causing changes in segregation ratios might be significant in evolution. Alleles, even if the most favourable, will not be perpetuated if the chromosomes bearing them are systematically excluded either from gametes or from zygotes. The importance of such factors depend on their extent in living organisms and this, at present, is largely unknown.

MUTATION

In chapter 3 we have already discussed the origin and also the different types and different rates of mutations which can occur. The Hardy-Weinberg equilibrium requires that mutation either does not occur, or that a mutation in one direction in one member of a population is balanced by a reverse mutation in another member. There are three possible ways in which mutations could change the gene frequencies in populations – directed mutation, random mutation, the provision of the raw material for selection.

Directed mutation

Mutation might be highly directed at a particular locus so that allele a^1 is consistently converted into allele a^2. The best evidence for directed mutation of this sort comes from work on maize. It is known that a mutator gene, Dt, can act on an unlinked gene, a, making it mutate to the A_1 form. This A_1 allele can order the production of the purple pigment anthocyanin in grains, stems and leaves. The same type of effect may be produced by the transposable elements we have already discussed, and these elements seem to migrate most often to 'hot spots' and therefore to affect particular genes.

Although these mutator genes and transposable elements may bring about directed mutation in specific genes, their effects on an evolutionary change in gene frequencies are not yet clear.

Random mutation

Mutation might be random but, with time, one allele may come to predominate over its alternative allele. This **mutation pressure**, if acting alone, is likely to require an enormous time to change the gene frequencies in a population to a significant extent. As you can see in figure 3.14 the rate of mutation varies with the species and with the actual gene. Prokaryote genes in general have a mutation rate which is lower than the 1×10^{-5} or 1×10^{-6} found in eukaryote genes. Imagine a very large eukaryote population in which the allele a_1 mutates to the allele a_2 at a rate of 1×10^{-6}, and that the population (to begin with) has a frequency of allele a_1 (p) of 0.9. If 1 in each million a_1 alleles mutates to a_2 in each generation it can be shown, although the mathematics is complicated, that after 1000 generations p will have changed from 0.9 to 0.899. If the generation time is very short this might be significant but for most organisms such a rate of change is too slow to matter, particularly as reverse mutation from allele a_2 to allele a_1 might also be occurring.

Provision of the raw material for selection

Mutation might provide the raw material, the **mutational currency**, on which selection may work. Many mutations, by upsetting a finely balanced metabolic system, may make the organism carrying them less likely to survive and so give a selective disadvantage. Less frequently a mutation may make the organism better equipped to deal with its environment and so give a selective advantage. It might seem from this preponderance of disadvantageous over advantageous mutations that a high mutation rate would be bad and would itself be selected against. There is some evidence that this is not so.

Two strains of the bacterium *E. coli*, in which recombination does not occur, were allowed to compete in laboratory culture. One strain carried a mutator gene which, unlike that in maize which we have already mentioned, was non-specific in its effects but greatly increased the rate of mutation of all the genes. The high mutation strain was always in low frequency at the beginning of the experiment but in every case it increased in frequency until it replaced the wild type strain. In the artificial conditions of the culture there might be many mutations which could increase the survival chance of the bacteria. These are more likely to occur in the mutator strain than in the wild type strain. So although large numbers of the mutator strain may be wiped out because

they have deleterious mutations, some may have advantageous mutations. These will survive better and leave more offspring than the wild
type strain. Eventually all the population will contain the advantageous
alleles – they will have reached fixation. Of course, in the absence of
recombination, the mutator gene itself will also have reached fixation. It
is as if the mutator gene, which has no phenotypic effect, has 'hitched a
ride' with the advantageous mutant allele. It is easy to see why this
whole concept has been called the '**hitch-hiker**' theory.

Neutral mutation

Many mutations may occur which do not seem to affect the phenotype
and these neutral mutations, which give neither an advantage nor a
disadvantage to the organism, would not be susceptible to selection.
Mutations may be neutral in effect because they occur in DNA which is
not transcribed, or in highly repetitive DNA. They may also occur in
DNA which is transcribed and translated into protein but, because
of the redundant nature of the genetic code, a change in a nucleotide
in the DNA does not always alter the amino acid sequence of the protein. As you have seen in chapter 3, a change in the third nucleotide
of a mRNA codon does not usually affect the amino acid specified –
GUU, GUC, GUA and GUG all code for valine, and GAA and
GAG code for glutamic acid (see figure 3.5). So the protein molecules
formed at translation would be unchanged. However, there is some
evidence that such **synonymous mutations** (mutations specifying the
same amino acid) might not be completely neutral in their effects. Translation rates may be dependent on the particular mRNA codons present
and so, although the protein itself is not changed, the rate of its production
may be altered in some way which could affect a cell's metabolism.

   Mutations which alter the amino acid sequence of a protein, such as
the change from normal to sickling haemoglobin, may have obvious
phenotypic effects and so be available for selection. Many amino acid
changes found by electrophoresis, however, have not yet been shown to
affect the phenotype and so must be considered to result from neutral
mutations. If a mutation is really neutral and selection has no effect on
it, then its fate is purely indeterminate. It might be eliminated as soon as
it appears, or it might move towards fixation in a seemingly haphazard
fashion called a **random walk**. Such a random change in the gene
frequency of a population is sometimes called **non-Darwinian evolution** because natural selection seems to play no role.

Mutation may alter gene frequencies in a population in many ways, but particularly by providing the 'new' alleles which may be tested by selection.

AN OPEN POPULATION – MIGRATION PRESSURE AND GENE FLOW

The genetic equilibrium of a population may be altered by the introduction of 'new' alleles. Sometimes the alleles are generated internally within the population by the process of mutation but often they enter from other populations which may be adjacent or further afield. Natural populations are open, unlike those which are investigated in population cages in the laboratory. In a deme there may be 30–50 per cent of newcomers in each generation. However, as the newcomers enter from adjacent demes which share rather similar environments this influx may have little effect on gene frequencies. It is when two populations are far enough apart to have different environments and different gene frequencies, but are also close enough for individuals to move from one population to the other that substantial changes in gene frequency may occur.

A relatively large influx provides a **migration pressure** which can influence gene frequencies in a particular direction. Migration may be random, as is shown in figure 5.1 with a group which is truly representative of the gene frequency of the original population moving away. If part of a population is not just colonizing a new, empty area by means of migration but is joining an already established population there will be a **gene flow** from the immigrants to the established population. There may be a change in the gene frequencies of the population joined by the immigrants but not of the population they left. Of course, if the random group moves away to colonize a new area and if it is substantial in size,

Figure 5.1 Migration between two populations

The migrants are truly representative of population A (light-coloured to dark-coloured moths in the ratio of 3 : 1). The migration has no effect on the gene frequencies in population A, but will affect those in population B where the original proportions of light-coloured to dark-coloured moths were different.

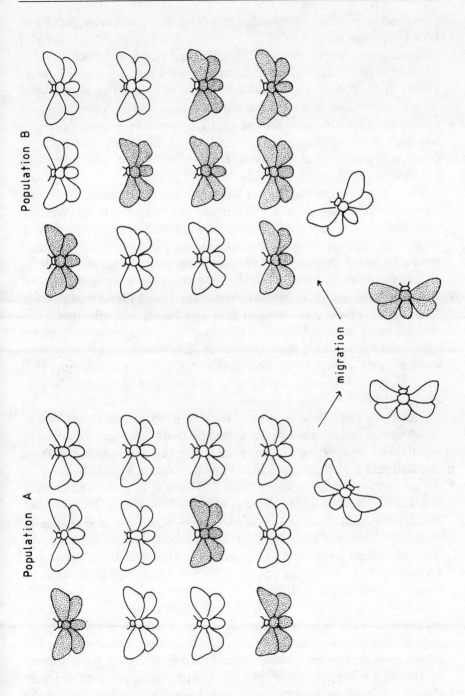

then the new and the old populations will be, at least initially, similar in gene frequencies.

Sometimes, however, the difference in behaviour which decides whether a particular animal will migrate or not is genetically determined. In the field vole, *Microtus agrestis*, some animals have genotypes which cause them to have high reproductive capacity but also to be intolerant of crowding. When the population numbers reach a particular level many of these density-intolerant individuals move out to colonize new areas, and so the gene frequencies of both the original and the new populations are altered.

The distance an animal moves from its birth place may also have an effect on gene frequencies. Little movement results in a small breeding unit with greater inbreeding and a subsequent steady loss in genetic variability. A great amount of movement may break social bonds and produce a loss of local adaptation. Very often emigration is not random, or genetically determined as in the field vole, but is strongly biassed with respect to age. As in man it is frequently young adults who go off to 'seek their fortune' elsewhere. It may also be biassed with respect to sex. In many animal societies, such as those of lions, young males almost invariably leave their families as they mature. They may then wander singly or with some other males until each can attach himself to another pride – usually by fighting and ousting the male which is already there.

The term gene flow is sometimes restricted to the result of the interbreeding which may occur along the boundaries of two adjacent populations which have distinct gene frequencies. If individuals from population X introduce more alleles into population Y than the reverse, the gene flow would be from X to Y. Often it is males rather than females which temporarily move across the boundaries. The result of this type of gene flow can be seen in studies of gene frequencies in populations of black and white North Americans. The black Americans were unwilling emigrants from Africa during the slave trade and formed separate populations from the white Americans in the areas to which they were taken. However, over the generations there has been a gene flow into the black population from white males who had temporary liaisons with black females. The gene frequencies of the black populations now, after ten or so generations, differ from those found in Africa and are moving towards those found in the white populations.

Although when discussing the effects of migration pressure and gene flow in a population we concentrate on changes in the frequency of the alleles of a particular gene, it must be remembered that whole organisms or whole gametes move into the population and not single gene loci. Other genes which enter may modify the effects of the gene being considered.

POPULATION SIZE AND NON-RANDOM MATING

The Hardy-Weinberg equilibrium requires a very large population throughout which there is the chance of random mating. Organisms in real populations, however, are generally much more restricted in their mating patterns. It may be that although mating appears to be at random it really occurs only between population members within a very limited geographical area. So demes, with their own patterns of gene frequency, are established within the large population.

It is also possible that mating is non-random for other than geographic reasons. Many organisms show **assortative mating patterns** in which they mate with other organisms of a particular type. In certain cases the matings may be between individuals which are more closely related than the average of the population to which they belong. This is sometimes called positive genotype assortative mating, but a much simpler term is **inbreeding**. The most intense form of inbreeding is self-fertilization. In this case the gametes are formed from a single parental genotype and the gene pool from which they are derived is therefore restricted to that one organism. Any type of inbreeding will lead towards homozygosity for many loci but self-fertilization will bring this about most quickly. See figure 5.2.

Self-fertilization occurs in a limited number of plant species, but in no higher animals. Homozygosity in higher animals is achieved by matings between close relatives, a situation which occurs more frequently in laboratory than natural populations. In self-fertilization the organism must mate with one of identical genotype, that is with itself.

Often an animal will mate with another which resembles it in some phenotypic way. Humans tend to match for size and colour, and also for intelligence level. However, even though animals may choose to mate with those whose similarity suggests some relationship, they usually do not mate in natural conditions with very close relatives. This can be avoided if they can recognize their relatives and so refrain from breeding with them. Often the outbreeding is achieved by the dispersal of the family, or maybe only of the males, before sexual maturity is reached.

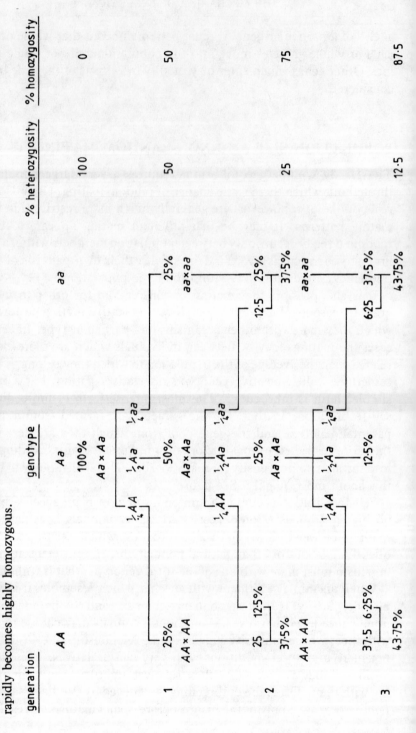

Figure 5.2 Expected increase in homozygosity due to self-fertilization

In each generation of self-fertilization the percentage of heterozygosity drops, so even a completely heterozygous population rapidly becomes highly homozygous.

In higher plants various methods for ensuring cross-pollination have evolved, often alongside the evolution of a particular insect vector, and some of the devices are extremely complicated. Sometimes cross-pollination may be achieved by simpler methods such as a difference in timing for the development of male and female parts of the flower. Pollination within the same flower can be totally avoided if the flowers are unisexual. Sometimes the male and female flowers are some distance apart on the same plant (monoecy, as in hazel) but sometimes there are separate male and female plants (dioecy, as in holly).

Many types of assortative mating patterns result in the effective breeding population being much smaller than might at first appear; the need for phenotypic matching, for example, makes a large proportion of the population irrelevant for mate selection. However, breeding can occur across these boundaries and the gene pools of the small sections and of the total population continue to contain a range of alleles for each gene locus.

If the total population is small, however, the gene pool is restricted. Even within this small population not all the organisms might manage to reproduce – there may not be enough mates of the right sex to go around, and age differences might mean that they are not all ready to mate at the same time.

Gene frequencies may therefore change from generation to generation. Such changes which result from the luck of mating are called **random genetic drift** and they may lead to the fixation of alleles in a small population within a relatively short time. Genetic drift may be investigated both by means of actual gene samplings from real populations and with computer simulations.

Random numbers are used in the simulations. These are numbers of one or more digits in which each digit has the same probability of occurring, and where there is no correlation between successive digits or numbers. A gene locus *Aa* is investigated and matched to the random numbers. If the original population has 50 per cent *A* alleles, then the digits 0,1,2,3 and 4 are taken to represent *A* and the remaining digits represent *a*. If the proportion of *A* alleles alters, then the representative numbers are altered accordingly. A frequency of 40 per cent *A* would mean that only digits 0,1,2 and 3 now represent the *A* allele, and digit 4 moves into the *a* allele class. The program in figure 5.3 shows how an allele can move to fixation fairly rapidly. Random genetic drift occurs in populations of any size but it is only in small populations that gene frequencies may change in a significant way.

Figure 5.3 Computer simulation of random genetic drift using random numbers

The program is written in BBC basic and deals with a population in which the original frequencies are 50 per cent *A* alleles and 50 per cent *a* alleles. An explanation is given in the text of the rule for transforming the randomly generated numbers into alleles – a transformation which depends on the gene frequency of the previous generation. As can be seen in the typical outputs given, an allele can move to fixation within a few generations in such a small population. Without such a weighting of alleles depending on their frequency within the parental generation, it could be expected that it would take 2^9 (512) generations of completely random change before one allele is lost. (Once the first allele of the 10 is fixed, there is a cumulative chance of 2^9 of getting a similar allele.)

Program

```
>LIST
   10 C=50:G=0:DIM A$(10):PROChead
   20 REPEAT:PROCcalc:PROCprint
   30 UNTIL C<10 OR C>90
   40 STOP
   50 DEFPROCcalc
   60 A=C:C=0:FORI=1 TO 10:P=RND(10)-1
   70 IF P>=(A/10) GOTO 90
   80 A$(I)="A":C=C+10:GOTO 100
   90 A$(I)="a"
  100 NEXT
  110 G=G+1
  120 ENDPROC
  130 DEFPROCprint
  140 FORI=1TO10:PRINTA$(I);:NEXTI
  150 PRINTTAB(17);C;TAB(34);G
  160 ENDPROC
  170 DEFPROChead
  180 CLS:PRINT:PRINT"Alleles";
  190 PRINTTAB(13);"Percentage A";
  200 PRINTTAB(29);"Generation"
  210 FORI=1TO39:PRINT"=";:NEXTI:PRINT
  220 ENDPROC
>
```

Sample printouts

```
RUN

Alleles        Percentage A     Generation
==========================================
aAAaaAAAaa     50               1
aaaaAAaaAa     30               2
AaaaaaaAAA     40               3
AaAAAAAAAa     80               4
AAAAAAAAAa     90               5
AAAAAAAAAA     100              6

STOP at line 40
>
```

```
RUN

Alleles        Percentage A     Generation
==========================================
aaAAaaAAAa     50               1
AAAaaAAAAA     80               2
aAAaAAaAAA     70               3
aAaAAAAAaA     70               4
AAAaAaAAAA     80               5
AAaAAAAaAA     80               6
AaAAaAAaAA     70               7
AAAAAAAAAa     90               8
AAAAAAAAAA     100              9

STOP at line 40
>
```

```
RUN

Alleles        Percentage A     Generation
=========================================
aaAAaAAAaa         50               1
aaaAaaaAaA         30               2
AAaaAaaaAa         40               3
aAaaAaaaaA         30               4
AaaaaaaAaa         20               5
aaaaaaaaaa          0               6

STOP at line 40
>
```

```
RUN

Alleles        Percentage A     Generation
=========================================
AAaaAAaAAa         60               1
AaaaaaaaAa         20               2
AAAaAAaaaa         50               3
AaaaaAAaAA         50               4
AAaaaaAAaA         50               5
aaAaaAaaAa         30               6
aaaaaaaaaA         10               7
aaaaaaaaaa          0               8

STOP at line 40
>
```

Very small populations of a permanent kind are not usually found in natural conditions – they either get wiped out or they manage to increase in size. Small populations of a temporary kind may result from many different factors. Sometimes there are obvious environmental effects. The cold British winters of 1961–2 and 1962–3 reduced the population of the Dartford Warbler, *Sylvia undata*, from 450 estimated pairs to ten recorded pairs, but the numbers recovered during the following ten years. In many organisms an originally small population may suddenly increase in size because of an improvement in environmental conditions or the removal of a predator. The population may then destroy its environment because of its own increased numbers and crash just as suddenly as it had earlier increased. It may be that certain of its members are more likely to survive through the population crash, and so the gene frequency will change during and after this genetic bottleneck of a very small population. However, even alleles which are advantageous for survival might be eliminated from the population if the very few members bearing them are wiped out by some external chance event. Whether or not a beetle is well camouflaged against its background is of no value if it gets squashed under the large wellington boot of the gardener.

As we have already said in chapter 2, small populations may also be found if a few organisms can manage to reach a new, empty habitat either as a result of some accident which carried them away from the main population or by limited emigration. The founding individuals have a very limited gene pool which may not always be representative of the population from which they moved. Think of an original population in which half of the members contain an advantageous allele, and from which two individuals are carried away to a new area. There is a probability of one in four that neither of the two individuals will contain the advantageous allele. If all the genes are considered it can be seen that there is a high probability that the founding individuals will have very different gene frequencies from the original population. As the founding population is initially small it is also possible for alleles to be lost by the process of random genetic drift. Add to this the fact that the new environment is likely to differ from the old, putting a premium on alleles which were not particularly advantageous in the environment of the main population and you can see that the organisms which do manage to survive in the new habitat are likely to show a rapid change in gene frequencies in their increasing populations – like Darwin's finches in the Galapagos Islands.

A species may sometimes rapidly extend its range because the few individuals which move away may find a new empty environment in which they flourish. The Collared Dove, *Streptopelia decaocto*, which is now a common bird throughout western Europe and the British Isles, extended its range from the Danube to Holland in 1947, to Belgium in 1952, to Norfolk in 1955, and reached Morayshire and Somerset in 1957. Such a rapid spread into previously uncolonized areas would suggest that it did not happen purely by chance, and Mayr thinks that a genetic change must have occurred which produced a greater tolerance to lower temperatures and so enabled the colonists to survive.

So changes in gene frequencies may be brought about because of assortative mating patterns or because of chance events in very small populations.

SELECTION

As you have seen, there are various mechanisms which can bring about changes in the gene frequencies found in populations. Meiotic drive, or differences in the viability of gametes; the occurrence of spontaneous mutation; emigration or immigration leading to a loss or gain of alleles; non-random mating, or the statistical effects of random mating in very small populations – all may have their effects. The effects, however, are subject to the fact that organisms live in actual environments and that some organisms are more likely than others to survive through their developmental stages and to achieve reproductive success. So **natural selection** must be considered to be the most important agent leading to substantial changes in gene frequencies in populations.

Each phenotype in a population usually has to compete with other phenotypes for reproductive success. Such success is determined by multiple interactions between the organism and its enemies, or its competitors, or with pathogens of various types, and also by the way in which the organism might cope with the non-biological environment which is never constant. Natural selection differentiates between phenotypes in a population with respect to their ability to produce offspring, and so determines their contribution to the gene pool of subsequent generations.

Mathematical concepts

In considering natural selection three terms to which mathematical values may be given are particularly important. These are:

1 **survival rate;**
2 **relative fitness;**
3 **selection coefficient.**

It is easier to understand the meaning of the terms by considering how they are calculated in a particular population. Let us take a hypothetical population in which we shall look at one gene locus with two alleles A and a, and let us assume that there is codominance so that the three genotypes AA, Aa and aa result in three different phenotypes on which selection can work. The numbers of the three phenotypes are counted or estimated (depending on the size of the population) both before and after a selective event, such as a change in temperature, occurs.

Numbers of genotypes

	AA	Aa	aa
Before selection	4000	5000	2000
After selection	3800	4000	1100

It is obvious that selection has reduced the numbers of all three phenotype classes, but not to the same extent.

The survival rate for AA is $\dfrac{3800}{4000} = 0.95$

The survival rate for Aa is $\dfrac{4000}{5000} = 0.8$

The survival rate for aa is $\dfrac{1100}{2000} = 0.55$

So the AA phenotypes are surviving best, and this genotype is defined as being the fittest. To find the relative fitness (symbolized by W) the survival rate of the fittest genotype is taken as the standard against which the others are compared.

Relative fitness of AA is $\dfrac{0.95}{0.95} = 1$

Relative fitness of Aa is $\dfrac{0.8}{0.95} = 0.84$

Relative fitness of aa is $\dfrac{0.55}{0.95} = 0.58$

The selection coefficient (symbolized by s) is then easily found as it is $1 - W$.

Selection coefficient of AA is $1 - 1 = 0$

Selection coefficient of Aa is $1 - 0.84 = 0.16$

Selection coefficient of aa is $1 - 0.58 = 0.42$

The relative fitness, W, reflects the chances of an organism's reproductive success; the selection coefficient, s, reflects the chances of its reproductive failure due to selection.

Although data can be collected within a single generation in laboratory conditions where the selective event may be controlled, it is more likely that the data collected from populations in natural conditions will be from two successive generations. In this case the calculations must be altered.

Relative fitness can be established by comparing the observed phenotype frequencies in the second generation with those which would be expected if the population was behaving according to the Hardy-Weinberg equilibrium and which would, therefore, be subjected to no selection at all. Again consider some figures from a hypothetical population, in which selection occurs in the second generation.

	Numbers of genotypes			Total number
	AA	Aa	aa	
Before selection – 1st gen.	3000	4000	2000	9000
After selection – 2nd gen.	3500	4250	1250	9000

The allele frequencies in the first generation can be calculated from the genotypes.

Frequency of allele A is $\dfrac{2(3000) + 4000}{2(9000)} = 0.55 = p$

As $p + q = 1$, q must be 0.45. This is the frequency of allele a.

The expected genotype frequencies in the second generation in the absence of selection may now be calculated.

Expected frequency of AA is $p^2 = (0.55)^2 = 0.3025$

Expected frequency of Aa is $2pq = 2 \times 0.55 \times 0.45 = 0.495$

Expected frequency of aa is $q^2 = (0.45)^2 = 0.2025$

The observed frequencies may be obtained from the data.

Observed frequency of AA is $\dfrac{3500}{9000} = 0.39$

Observed frequency of Aa is $\dfrac{4250}{9000} = 0.47$

Observed frequency of aa is $\dfrac{1250}{9000} = 0.14$

A figure analogous to the survival rate can be taken as the ratio of the observed to the expected frequencies.

Ratio of observed/expected frequency for AA is $\dfrac{0.39}{0.3025} = 1.28$

Ratio of observed/expected frequency for Aa is $\dfrac{0.47}{0.495} = 0.95$

Ratio of observed/expected frequency for aa is $\dfrac{0.14}{0.2025} = 0.69$

Relative fitnesses and selection coefficients may now be calculated as before.

Relative fitness of genotype AA is $\dfrac{1.28}{1.28} = 1$

Selection coefficient $= 0$

Relative fitness of genotype Aa is $\dfrac{0.95}{1.28} = 0.74$

Selection coefficient $= 0.26$

Relative fitness of genotype aa is $\dfrac{0.69}{1.28} = 0.54$

Selection coefficient $= 0.46$

Modifiers; linkage disequilibrium; supergenes

For simplicity we have been considering a single gene locus of a particular type (two alleles with codominance) in these calculations but generally selection acts on the whole organism, not on the phenotypic effects of single genes. Even in cases where a single locus is of very great importance as in the industrial melanism shown by the Peppered Moth, where selection seems to act on black or light colour according to the background, there is evidence that **modifier genes** which have altered the dominant phenotype are also involved.

There may also be modifier genes present in the genome of an organism which seem to have no phenotypic effect. The *sepia* eye allele is found in natural populations of *Drosophila melanogaster* in very low frequency, and this suggests that it is constantly selected against, but is maintained by mutation pressure. If *sepia* eye bearing flies are placed in a population cage with flies from a highly inbred laboratory strain, the *sepia* eye allele usually increases in frequency and eventually stabilizes at a higher level than in the wild. So the selection pressures in the population cage seem to be measurably different from those in the wild. However, if the experiment is repeated using a different batch of wild-type flies, the *sepia* eye allele again stabilizes at a higher than natural frequency but the level is different from that reached in the first experiment. A further batch of wild-type flies leads to stabilization at a third level. The three batches of wild-type flies seem differentially to affect the selection pattern against the *sepia* eye allele under laboratory conditions and this suggests that they contain different frequencies of fitness modifier genes.

One of the main ways in which genetic variation is produced in populations is by crossing-over in meiosis. If a particularly advantageous set of alleles came together because of recombination, it would be useful if the set were maintained and preserved by cutting down any crossing-over which could separate the advantageous alleles. This non-random assortment of alleles on chromosomes is called **linkage disequilibrium**, and there is evidence that selection for whole sets of alleles does occur in some species.

Females of several species of Swallowtail butterflies show **Batesian mimicry** – they gain some protection from predators by resembling other species which are distasteful. Clarke and Sheppard in 1971 showed that the mimicry patterns in *Papilio memnon*, an East Indies species of Swallowtail, could be explained as being controlled by five

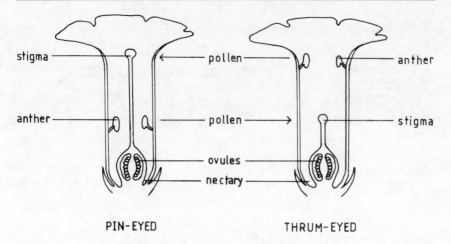

Figure 5.4   Heterostyly in primrose, *Primula vulgaris*
The flowers are adapted to be cross pollinated by a visiting insect, and show
self-incompatibility for pollen growth as well as other adaptations.

closely linked loci which determine the phenotypic effects of the
presence or absence of tails, the forewing pattern, the hind wing
pattern, the colour of the basal triangle of the forewing, and the colour
of the abdomen. It is important that a mimic should strongly resemble
the model, otherwise the whole point of mimicry would be lost. A
mimic which looks like the model in some ways but not in others will
not survive as well as one which has 'got its act together' and looks very
like the model. In terms of television acts, it is not enough to wear a red
fez to imitate the late and great comedian Tommy Cooper. Without the
arm movements, the look of bewilderment, and the 'just like that' the
impression is not good enough to persuade us. So it is with the Swallowtail – all the patterns must be right. The mimic patterns behave as if
they are controlled by a single gene, and the five linked loci are called a
supergene.

Other examples of supergenes are known. In the primrose, *Primula
vulgaris*, at least seven gene loci are involved in the supergene for
heterostyly. They determine such variable traits in the style of the
flowers as the area of conducting tissue for pollen tube growth, and
the style length. They also determine the position of the anthers and the
size of pollen grains. The supergene has two superalleles, S and s, which
control whether the plant is thrum-eyed or pin-eyed.

A

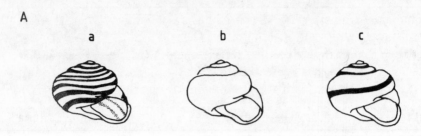

a   b   c

B

Figure 5.5   Shell polymorphism in *Cepaea* species
A   *Cepaea nemoralis*
       a 5 banded form   b unbanded form
       c mid-banded form
B   *Cepaea hortensis*, 5 banded form

Pin-eyed plants (*ss*) have flowers with high stigmas and low anthers; thrum-eyed plants (*Ss*) have low stigmas and high anthers. As you can see in figure 5.4, a visiting insect, such as a moth which has a proboscis which is long enough to reach the nectary at the base of the flower, will accidentally pick up pollen at a particular proboscis level in one type of flower, and transfer it to the stigma in the other flower type. Thrum pollen, even though it can germinate on a thrum stigma, fails to penetrate the stigmatic surface. So any pollen which accidentally falls from the high anthers in a thrum plant down onto the stigma cannot bring about self-fertilization. Recombination is rare within the supergene and so most plants are either pin-eyed or thrum-eyed, and must outbreed with plants of the other type.

The patterning on the shell of the European land snail, *Cepaea nemoralis* is also controlled by a supergene. Some individuals have plain shells which may be brown, pink or yellow, whereas others have a variable number of dark brown bands over the top of this ground colour. At least four loci are involved which determine the ground colour, the number of the bands, the depth of colour of the bands and the amount of spread of the bands.

A vast amount of investigation has been carried out on *Cepaea nemoralis* and it is known that very many evolutionary mechanisms can affect the gene frequencies in different populations. There is clear **visual selection** by predators – certain shell patterns seem to provide camouflage against particular backgrounds. However, there are difficulties. There is another common species of *Cepaea*, *Cepaea hortensis*, which also has variable shell patterning and which may live in the same locality as *Cepaea nemoralis*. Even though the populations of the two species may live against the same background and be hunted by the same predators, their gene frequencies may be very different. In beech woods, for example, *Cepaea nemoralis* populations consist almost entirely of unbanded browns whereas *Cepaea hortensis* populations are mostly made up of heavily banded yellows.

The different colours of the shells may have an effect by determining the energy uptake of the snails from sunshine, and there is experimental evidence that **climatic selection** may be important. In many places, however, an area in which there is a constant allele frequency is separated by a steep cline from adjacent areas with very different allele frequencies, even though the transition does not coincide with any abrupt temperature, or other environmental, change.

It is also possible that **random genetic drift** could be important in certain small populations of *Cepaea*. When this is added to the effects of selection by differential predation or by different climate, or to **area effects**, it becomes obvious that there is no simple answer to explain the evolution of the various *Cepaea* populations.

Hard and soft selection

Sexually reproduced organisms being genetically unique, and the environment in which they live being very variable in space and time, it is obvious that natural selection must be regarded as a very changeable force. Strong selection against the phenotype produced by a particular allele, superallele or genome at one time or in one place might change to selection in its favour either shortly afterwards or nearby if the environment changes. Sometimes a particular genotype is so unfitted to the usual environment that it will always be eliminated by selection, and this has been named **hard selection**. There are many examples of alleles which are lethal if homozygous, and their lethal effect is not ameliorated by the rest of the genome or by the environment. The *yellow* gene in the mouse, and the *tailless* gene in Manx cats are of this kind, and so all yellow mice and tailless Manx cats are heterozygous. At least twenty-seven recessive lethal alleles may be found in cattle, the best known of which results in 'bulldog' calves, so named because of their abnormally large heads with flattened faces and their short legs. The calves were first described in Germany in 1860, and they are generally aborted at about six to eight months. Many of the lethal genes in cattle have been spread over large areas because of the practice of artificial insemination.

The lethal effect may result in death almost as soon as the alleles from the gametes come together in the zygote. It may be delayed through a variable period of time to give an early or a late abortion in mammals, or the organism may die relatively soon after birth. Lethal alleles are also found in plants. A genotype which leads to a lack of chlorophyll production will mean the death of the seedling as soon as its store of food in the seed has been exhausted.

Hard selection against certain genotypes may also occur as a result of an aspect of the environment such as a particular temperature. Whether the selection is the result of such environmental stress or because of the presence of lethal alleles, it will lead to a reduction in the size of the population.

If, however, the population has reached the carrying capacity of the environment so that any 'extra' individuals must die, it is likely that those which are eliminated will be the less well adapted. Selection against these does not decrease the expected population size – the same number of 'extra' individuals would have died even if they were all equally adapted – and so this has been called **soft selection**.

Main types of selection

Natural selection is a process which reduces the proportion of organisms with low relative fitness within a population and which, therefore, must increase the proportion of organisms which are nearer the optimal phenotype for the specific environmental and competitive conditions. A high survival rate is of importance in that it gives the organism a higher chance of reproducing and therefore making a contribution of its genes to the gene pool. Most phenotypic traits are controlled by many genes and tend to show a normal distribution in the population. The position of the optimal phenotype will determine the basic type of selection.

As you can see in figure 5.6, there are three basic types of selection which result in near-optimization (getting the best result) for the population in a particular set of conditions.

1 **Stabilizing natural selection** – this favours the mean at the expense of the ends of the distribution.

2 **Directional natural selection** – this favours one extreme of the phenotypic range.
3 **Disruptive natural selection** – this favours both extremes at the expense of the mean.

Although we can categorize selection in this way, we must remember that the classes are all aspects of the same process. The very fact that some organisms have a lower relative fitness (that is, they have lower survival and/or reproductive rates than others) means that those which keep the population going are competing with organisms which are

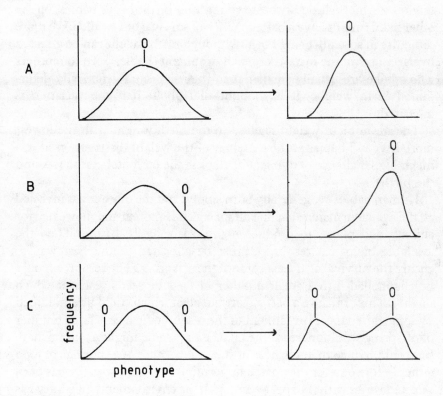

Figure 5.6   The three basic types of selection
A   stabilizing natural selection
B   directional natural selection
C   disruptive natural selection
The optimal phenotypes to which selection is acting are shown as 0.

becoming more and more similar to themselves. If directional natural selection gradually eliminates one extreme of the phenotypic range, then eventually there will be a random distribution around the optimum phenotype. Stabilizing natural selection will then tend to maintain this. Keeping this in mind, let us now consider some examples of these three types.

Stabilizing natural selection

This type of selection favours the mean as opposed to the extremes of the phenotypic range. Sometimes its mechanism seems reasonably clear. Given the range of possible birth weights in man, it is likely that very small babies would stress the mother less during childbirth but might have more difficulty in surviving after birth. Large babies, on the other hand, might have a better chance of survival but would have more difficulty in actually being born. Although birth weight can be affected by variations in the intra-uterine environment (babies born to mothers who smoke are usually smaller than their expected genetically determined birth weight) it is thought that about half the variation is inherited.

The mean birth weight is close to the birth weight with the lowest mortality, and babies at the extremes of the weight distribution have a much higher chance of dying at birth or in the perinatal period (around the birth).

Human babies are generally born singly, and the mother has invested all the available material in one offspring. It is more usual, however, for animals and plants to produce many offspring at a time. Then the available biological resources must be allocated in the best way to ensure the survival of at least some of the offspring. The resources could be channelled into a small number of eggs or seeds, each of which might, therefore, have a better start but all of which could be eliminated by some mischance. Investment of the resources in a very large number of offspring would increase the chance that some might avoid mishap, but could reduce their chances of survival because of a shortage of food either before or after they hatch or germinate. Which strategy has been selected varies with the species and with the environment. Salisbury has shown that plants which produce large numbers of tiny seeds generally live in open habitats such as fields or disturbed earth. Here there is a reasonable chance that at least some seeds could start to grow and be well established before the seed food store runs out. Seed size is larger in species which inhabit scrub and woodland margins, and becomes largest in woodland shrubs and trees where it may take much longer for the seedling to become self-sufficient.

The number of eggs laid by female birds is not directly related to the amount of food available to the hen – if eggs are removed from the clutch she will continue to lay until she reaches the visual pattern produced by a particular number (which is how we can get our domestic hens to go on laying day after day). It is, however, related to the number

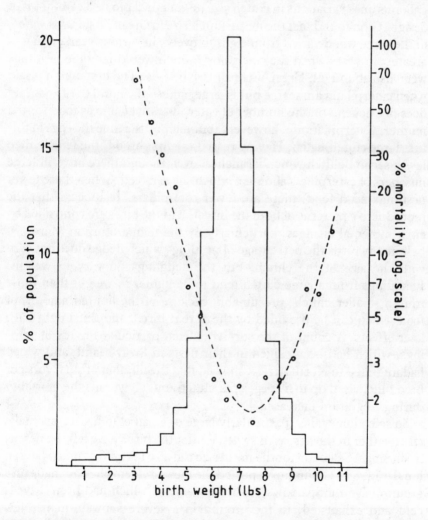

Figure 5.7 Distribution of birth weights of 13 730 children born in University College Hospital between 1935 and 1946 *From Mather*.

of offspring for which food can subsequently be found by the parents. The female English Swift, *Apus apus*, usually lays two or three eggs per clutch, but the percentage of nestlings which fledge is lower if there are three in the nest to be fed on collected insects than if there are two. Females which lay two eggs leave a greater number of progeny on average than those which lay three eggs (although these may be at an advantage in years when there is an excess of food available, such as the particularly fine summers of 1949, 1951 and 1955).

Sometimes variations in clutch size might appear to be a careful bit of design. Gibb found that the mean clutch size in Great Tits, *Parus major*, in the same wood varied from eight to twelve in various years, and was greatest in years when the caterpillars which would feed the nestlings were also abundant. Great Tits rear two broods. The first clutch is laid when caterpillars are scarce but is large, and so the number of eggs laid does not depend on the number of caterpillars available as food for the mother. Caterpillars are, however, abundant by the time the eggs in this first large clutch hatch. Having reared the first brood, the female then lays a second clutch which is smaller, even though there are still large numbers of caterpillars about on which she can feed. When these fewer nestlings need food, there are fewer caterpillars. It appears that the reproductive patterns of both the insects and the birds are controlled by environmental changes, particularly by the temperature in March.

In many birds the percentage of nestlings which fledge does not vary with the size of the clutch, but the fledglings do vary in weight. Fledglings from a large clutch tend to be lighter in weight than those from a smaller clutch, presumably because of the limited amount of food which can be provided by the parent birds, and this might have later effects. A count of the survivors from particular broods of starlings, which had been ringed in their nests in Switzerland, and which had migrated to North Africa, showed that the number of survivors per brood increased up to five eggs per clutch, but above this the mortality during migration increased.

So selection against particularly large or small clutch sizes may take effect either in the nest as in swifts or after the birds have left the nest as in starlings. Selection for the mean size can also take effect much later in life. In 1898 H C Bumpus reported the measurements he had made on a group of House Sparrows, *Passer domesticus*, which had been driven, cold and exhausted, to the ground in a severe snowstorm in New England. Out of a total of 136 birds, seventy-two survived. He made nine measurements on each bird including body weight, total length and wing span, and concluded that it was the more variable birds which perished.

Sometimes stabilizing selection may occur at a particular gene locus because the heterozygote has higher relative fitness than either homozygote, and this is one of the ways in which deleterious alleles may be maintained in a population. It might be expected that a mutation such as that which affects human haemoglobin and produces sickle cell anaemia would be strongly selected against. People who are homozygous for the mutant allele may have many incapacitating symptoms (see figure 3.3) and are unlikely to survive long enough for them to reproduce. Heterozygotes have less than 1 per cent of their red blood cells deformed by the abnormal haemoglobin, and the presence of the mutant allele gives some protection against malaria. The falciparum malarial parasite, *Plasmodium falciparum*, is carried from person to person by certain mosquitoes and develops within the person's red blood cells. The sickling allele affects the blood cells so that they provide an unsuitable environment for the developing parasite. To be homozygous for the sickling allele might give even greater protection against malaria but the other effects of the homozygosity are such that a

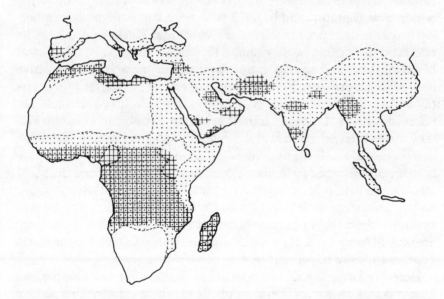

Figure 5.8   Distribution of the sickle cell allele (cross hatched) in areas where malaria was endemic (dotted)   *Based on Friedman and Trager*.

homozygous child is likely to die at an early age. So the heterozygote has a greater relative fitness than the homozygote for the sickling allele. In regions where malaria is endemic the other homozygote, with normal haemoglobin, also has low relative fitness because such a person will be susceptible to the disease.

The advantage to the individual of being heterozygous and therefore able to survive in malarial areas is at a cost to the population of the loss of the many homozygotes who will be born as a result of crossing two heterozygotes, a loss which also, of course, can affect gene frequencies at other loci. If the environment is radically altered by the elimination from an area of the mosquitoes which carry the malarial parasites, then the relative fitness of the normal phenotype will increase and the sickling allele will gradually be removed from the population by means of directional selection.

In this case, a change in the environment can lead from stabilizing to directional selection in favour of one homozygote. An environmental change can also, as we have mentioned, produce directional selection which becomes stabilizing selection favouring an optimal, mean phenotype. The rat poison Warfarin was introduced in 1950 and became the poison of choice mainly because of its low toxicity to other animals. By 1958 there were reports of rats which were resistant to the poison near Glasgow, and by 1972 twelve areas of Britain had temporary or permanent populations of resistant rats. Warfarin produces its effect by interacting with vitamin K which is necessary for normal blood clotting, and so may cause the death of the animal by massive internal haemorrhage. Resistance to the effects of Warfarin is determined by a single dominant gene, and rats which carry this allele can offset the effects of the interaction between the poison and vitamin K if they can get a larger amount of the vitamin in their food. Rats which are heterozygous need slightly more vitamin K but those which are homozygous for the resistance allele need up to twenty times the usual amount of vitamin K, a requirement which they cannot usually fulfil. So the homozygotes will have lower relative fitness than the heterozygotes – the resistant homozygotes because of the prohibitively large amount of vitamin K they need, and the non-resistant homozygotes because they are wiped out by the poison.

Heterozygotes may also be selected in a population because the two homozygotes have a combination of alleles which is lethal irrespective of environmental circumstances. Two dominant alleles, *curly* wings *Cy* and *plum* eye colour *Pm*, are linked on the second chromosome of

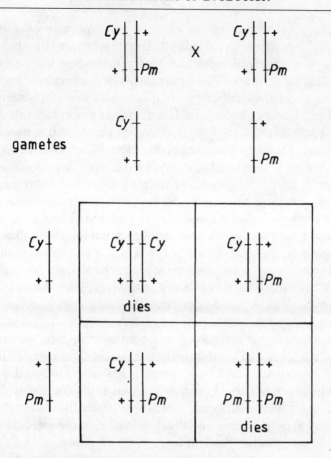

Figure 5.9 Balanced lethals in *Drosophila*
The two genes curly-wing, *Cy*, and plum-eye, *Pm*, are linked on the second chromosome, and each is lethal if homozygous. All the surviving offspring are heterozygous like their parents.

Drosophila, and crossing-over between them is prevented by a chromosome inversion. Both are lethal when homozygous, and so all individuals which show a curly winged, plum eyed phenotype are heterozygous. If two heterozygotes are crossed, half the progeny may die and the survivors are curly winged and plum eyed like the parents. The cross can be seen in figure 5.9.

The heterozygotes behave as if they were homozygous and breed true, and so such **balanced lethals** maintain a frequency of 0.5 for each allele for ever.

As you see, heterozygotes are often at an advantage over homozygotes and this can be seen particularly clearly when two highly inbred strains (which are likely to be homozygous at many gene loci because of the inbreeding) are crossed. The F_1 organisms are generally found to be hardier and larger than either parental type and this phenomenon is called **hybrid vigour** or **heterosis**. Heterosis has not been exploited in animal breeding to the same extent as in crop plants, but is now being used to increase meat, milk and egg production. When inbred Hereford bulls were mated to unrelated cows from a different line, the offspring weighed about 12 per cent more at weaning time than inbred calves. This hybrid advantage continued, the final weight of the hybrids being on average thirty kilograms more than the inbred animals.

Heterosis in crop plants has been demonstrated for a long time. W J Beal, working in the second half of the Nineteenth Century, showed that crossing different varieties of maize produced offspring which were superior to the inbred varieties. Work with crop plants has continued throughout this century and enormous increases in production have been obtained.

The genetic basis for heterosis is still controversial but two theories have been suggested. The **dominance hypothesis** suggests that in selecting for particular traits the breeder has also produced strains which are homozygous for certain deleterious recessive alleles. When two organisms from two different strains are crossed, it is likely that the offspring will acquire dominant alleles at most of these loci and therefore have fewer genetic disadvantages than its parents. The second theory, the **overdominance hypothesis**, suggests that the offspring, which is much more likely to be heterozygous at many loci than its parents, will show heterozygote advantage and will, because of this fact, be more vigorous. It is possible that the two enzymes which could be produced by a heterozygous cell could buffer the cell or the whole organism from environmental fluctuations.

These are examples of artificial selection to obtain improved yields, but heterosis also occurs in natural populations. Self-pollination in some plants will give rise to populations made up of individuals which are homozygous at many gene loci and so are genetically very similar to each other. Some organisms are parthenogenetic – they produce offspring without fertilization occurring. However, many of these self-fertilizing or parthenogenetic species may, at certain times, produce offspring by means of sexual out-crossing. When this happens, large numbers of new clones or pure lines which may be better than their parents may result, and selection can then pick out the best types for particular environments.

Directional natural selection

In this the optimal phenotype is at one end of the distribution and selection eliminates phenotypes at the other end. Many of the examples described in chapter 2 show this type of selection. Phenotypes of the Peppered Moth are selected according to their camouflage against tree trunks which may be blackened by industrial pollution. *Agrostis tenuis* on spoil heaps, or *Ectocarpus* on ships' hulls have been selected according to their tolerance to heavy metal ions.

Directional selection may also be seen in the spread of resistance to antibiotics in bacteria or resistance to insecticides in the insect vectors of pathogens (disease causing agents). In bacteria, resistance genes can spread by the transmission of the plasmids (which include the genes) from one bacterium to another. The very short generation time produces a resistant population very quickly. The same unfortunate effect is seen in insects. A survey of the mosquito *Anopheles gambiae* in villages in Nigeria which had not been sprayed with DDT showed that between 0.4 and 6 per cent of individuals were heterozygous for a DDT resistance gene. After spraying, 90 per cent of the mosquito population were homozygous for the resistance gene.

Resistance, in general, is produced by many genes. In House Flies, *Musca domestica*, there are at least three genes for DDT resistance and two for organo-phosphorus resistance. Two of the DDT loci are concerned with chemical detoxification of the DDT but work through different biochemical routes, and they interact with a third gene which reduces the permeability of the cuticle. Similarly many different genes are known to be involved in resistance in *Drosophila melanogaster*, and selection in favour of resistance can be demonstrated in the laboratory. A large number of flies are exposed to a particular concentration of DDT under carefully controlled conditions. The DDT concentration is chosen so that some, but not all, of the exposed flies die from its effects. The survivors are mated and the large numbers of offspring are then exposed to the same concentration of DDT. As this is continued through many generations, a decreasing proportion of each exposed batch of flies die. There is a gradual increase in the mean resistance of all the individuals in the population. Although the reshuffling of alleles in meiosis and zygote formation will produce a range of combinations (and therefore resistance levels) in the population, the gene pool within which the shuffling takes place will contain a higher proportion of resistance alleles in each generation. The way in which resistance is produced varies from one inbred strain to another. In one resistant strain, genes on a particular chromosome may be particularly effective in producing resistance with other resistance genes being less import-ant. In a different strain these other genes might be more effective. If two such different strains are allowed to mate, the resulting hybrid flies are likely to be even less resistant than the original parental stocks before selection started. Reshuffling to produce the hybrid flies des-troys the useful gene combinations which are responsible for resistance in each of the two parental strains. Reproductive isolation between the two strains will allow high levels of resistance to develop.

Reproductive isolation is also important in organisms which have been introduced, possibly by chance, into new areas and directional selection eliminates individuals which are less fit in the new environ-ment. The optimal phenotype will differ from place to place in one new island, or between one island and another, and adaptive radiation by means of directional selection may be rapid.

Disruptive natural selection

Here two (or more) phenotypes are selected with intermediates between them being discriminated against, the term being introduced by Mather. The effectiveness of such selection has been demonstrated in laboratory experiments by many investigators. In some experiments, physical characters, such as the number of bristles on the thorax of *Drosophila*, provided the basis for selection. Flies from the top end of the range of bristle number were mated amongst themselves; those from the lower end of the range were similarly mated amongst them-selves. Within a few generations of such selective mating, the two sub-populations began to show significant differences in their mean bristle number.

Other investigators selected for behavioural rather than physical characteristics. *Drosophila* populations show variation in the amount of spontaneous movement which occurs when each fly is placed in stan-dardized conditions so that its activity may be recorded. Connolly selected for the two extremes of such a spontaneous activity range in a laboratory population, mating the low activity, rather sluggish males and females together, and the hyper-active males and females from the other end of the range together. Figure 5.10 shows the results of such

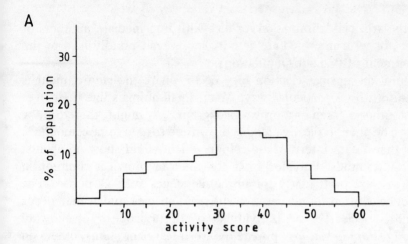

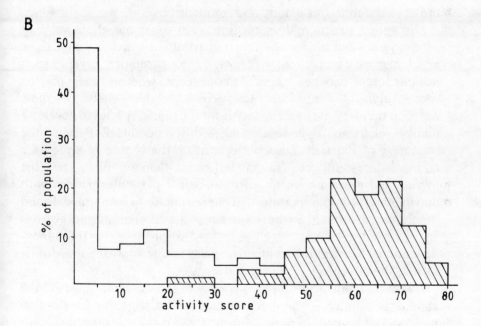

Figure 5.10 Drosophila spontaneous activity curves
A   frequency distribution of spontaneous activity in the original population
B   two distinct populations have emerged after selection for fourteen genera-
     tions.
From Connolly.

selection for fourteen generations. Two distinct populations have emerged.

So the artificial 'elimination' of flies with intermediate numbers of bristles, or with intermediate activity levels, can rapidly lead to the development of two sub-populations.

Natural disruptive selection may occur where the environment is non-uniform in a particular way. Many small animals live their lives against either a green or a brown background. It might, therefore, be better to be phenotypically green *or* brown and so gain some camouflage rather than be an intermediate colour which would show up against both backgrounds. Polymorphism for green or brown is common in larger insects, particularly for immobile stages such as pupae. The difference is often found between different species, but it may occur within a species. Some individuals of the European Swallowtail Butterfly, *Papilio macaon*, pupate on green leaves or stems, others on leaves or stems which are brown. Two distinct colour forms, green and brown, are found but almost no intermediates.

Disruptive selection does not always involve the elimination of heterozygotes but it does result in a distribution which is bimodal, or which may have more than two peaks. If the environment is very varied one phenotype may be successful in one niche, whereas others may be successful in different niches. In species which have evolved to show Batesian mimicry the best strategy is for the mimic species to develop a number of forms, each resembling a different model. Because the advantage of Batesian mimicry depends on the chance of a predator taking a distasteful prey (the model) rather than an edible prey (the mimic) and therefore learning to associate a particular pattern with unpalatability, it is important that there should be many models and few mimics in an area. So the frequency of the different mimics also has an effect and selection, as well as being disruptive, is also **frequency-dependent**. If the numbers of a particular mimic approach those of its model, the advantage of mimicry decreases.

It is also advantageous for prey species to show polymorphism in the absence of mimicry – the predator may form a searching image of its prey, and several different prey forms would make this more difficult.

Genetic isolation may result if the organisms of each particular optimal phenotype tend to mate with others like themselves. The population may then split into isolated breeding units. This all depends on the persistence of the varied environmental conditions to which the different types are adapted.

Variation from the main types

We have been discussing the basic forms of selection towards the optimal phenotype(s) for a particular set of environmental circumstances but you can see that in certain instances like mimicry, selection, although disruptive, is also frequency-dependent. There are many variations from the basic forms of selection which we must now consider.

Frequency-dependent selection
There are many instances where a phenotype is favoured if it is rare. The result of such successful frequency-dependent selection is, of course, that the rarer phenotype becomes commoner and the advantage it gains purely from rarity is lost. If the only advantage the particular phenotype displays is its rarity, it will not necessarily maintain this higher frequency and will tend to be selected against. The final frequency will depend on the balance between its selective advantage because of rarity and its selective disadvantage if it is not an optimal phenotype for the environmental circumstances.

We have already mentioned the advantage of a low frequency of mimics compared with the frequency of their models. Many investigators have studied the effects of altering the frequency of different types of prey offered to predators. Popham experimentally studied the relationship between the Rudd, *Scardinius erythrophthalmus*, (a fresh water fish) and its prey, the Corixid bug, *Sigara distincta*. The bugs may be various shades of brown, giving varied degrees of camouflage against the bottom of the lake or stream. Popham placed known numbers of two different shades, such as light brown and dark brown, in a tank which gave a uniform background and then counted how many of each type were eaten by the Rudd in the tank. In each case, as expected, he found that more of the least cryptic (less well camouflaged) bugs were eaten by the fish. However, both types were taken more frequently than expected when common and less frequently than expected when rare. Rarity conferred an advantage on both phenotypes whether they were well camouflaged or not.

All predators have to spend a certain amount of time and to expend a certain amount of energy in finding and catching their prey, and so would tend to take the commoner prey animals. They may have formed a searching image, which is that of the form they meet most often, and they would then ignore rarer forms. They may ignore novel stimuli rather than waste time and energy in investigating something which may turn out not to be a food source. So they would ignore the rarer forms which they meet infrequently. Allen and Clarke studied the reaction of British garden birds to coloured artificial 'caterpillars' made of a mixture of flour, fat and food colouring (see experiments at the end).

Many species of animals need to search for a mate as well as for food, and frequency-dependent selection may also apply here. In the Scarlet Tiger Moth, *Panaxia dominula*, each of the three genotypes at a particular locus has its own phenotype.

Mating seems to take place between unlike phenotypes, apparently because the females reject males of similar phenotype to themselves. Less frequent phenotypes are therefore more likely to mate as most of the available partners will be of the commoner type.

In some plants self-incompatibility means that out-breeding must occur. Many species possess a genetic locus, the self-incompatibility locus, at which there are a series of alleles. Pollen which contains the same allele as the stigma plant either cannot germinate or cannot make sufficient growth in the style if it does germinate. If there are three alleles s_1, s_2 and s_3 at the locus, all at equal frequency, then pollen may be able to germinate in only a third of the recipient plants.

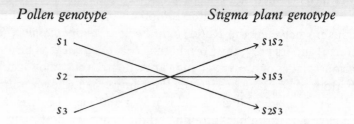

If a new allele, s_4, arose by mutation it would have an immediate advantage in that s_4 pollen would function in any of the recipient plants. So the s_4 allele would increase in the population and its selective advantage due to rarity would decrease.

Density-dependent selection

Density-dependent factors often involve **competition** for resources either within a population or between different populations. Such competition may be shown clearly by static organisms like plants where zones of resource depletion may be created around high density groups

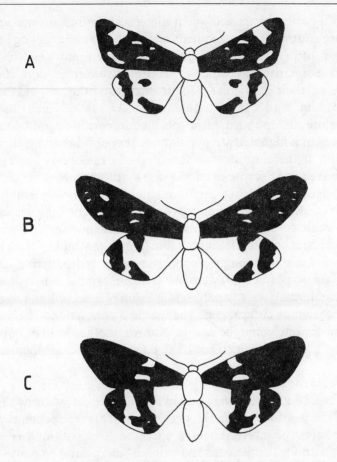

Figure 5.11   Polymorphic forms of *Panaxia dominula*, the Scarlet Tiger moth
A   the normal homozygote
B   the heterozygote, *medionigra*
C   the rare homozygote, *bimacula*
Drawn from photographs in Ford.

– the plants quickly use up the minerals available in the soil. Each plant in such a group will make proportionately less growth and produce proportionately fewer seeds than if it were in a group at lower density. A single plant of Fat Hen, *Chenopodium album*, may, if highly crowded, develop as an unbranched individual five to ten centimetres high bearing one to four seeds. Isolated from neighbours and well provided with nutrients, it may grow into a much branched plant 100–150 centimetres high which can bear 100 000 seeds. A particular resource level will support a particular mass of plant material, whether the mass is made up of large numbers of small plants or a smaller number of large plants – a concept formalized as the **Law of Constant Final Yield**.

If some organisms are able to use limited resources more efficiently because of their particular genetic make-up, they would be at an advantage in high density populations. An allele which makes the bearer more susceptible to intra-specific competition would decrease in frequency in a high density population. It could, however, increase in frequency if the restraints on density were removed. The restraints might be relaxed because of an increase in the available amount of food or a change in some other environmental factor. Some members of an animal population may emigrate when the population level is such that intra-specific competition with its subsequent shortage of food and its hormonal stress makes life intolerable or insupportable. Often population levels may be brought down by relative reproductive failure of certain genotypes at high density. Genetic variation in reproductive rates or efficiency, or a variable emigration rate according to genotype may bring about cyclical gene changes in a population. There may be selection for one genotype at low, and for another at high population density – a situation which occurs in many species of small mammals.

r and K *selection*

Sometimes competition may be obvious and direct, with two individuals fighting for food, or mates, or territory. This type of competition is called **contest** or **interference** competition. Competition may also be indirect with the competitors showing no behavioural response to each other, but both using up the limited resources. The winners are those which acquire the biggest share of the resources and so this is called **scramble** or **exploitation** competition. The ability to be a good contest competitor in a high density population or to be a good scramble competitor and so a good colonizer of new ground is the result of many complex factors. Each could give a selective advantage under particular and different environmental conditions.

Some individuals and species have been selected for their colonizing ability – their ability to move into a new geographic or ecological space not previously inhabited by the species. Assuming that this new space has untouched resources, the population will begin to grow exponentially. The **intrinsic rate of increase** of an individual is symbolized by r. Those individuals which have the highest rate of increase under these conditions will have the highest fitness. So selection will be for a high r

	r selection	K selection
Habitat/climate	temporary, variable	stable, fairly constant
Mortality	often catastrophic density independent	density dependent
Population size	variable in time usually well below carrying capacity of the environment recolonization each year	fairly constant in time at or near carrying capacity of the environment no recolonization necessary
Competition	variable	usually keen
Selection favours	1 rapid development 2 high maximal rate of increase (r) 3 early reproduction 4 small body size 5 single reproduction	1 slower development 2 greater competitive ability 3 delayed reproduction 4 larger body size 5 repeated reproduction
Length of life	short, usually less than 1 year	longer, usually more than 1 year

Figure 5.12 Table to show some aspects of r and K selection *Taken from Pianka.*

value. Of course, resources will decline as the population grows in size, and a new colonization then occurs.

Any species which is adapted to live in a short-lasting, unpredictable environment such as a new clearing in a forest will succeed if it can discover the habitat quickly, and then reproduce rapidly to use up as much of the available resources as possible before other species enter or the habitat disappears. It must then be able to disperse in search of new habitats as the original one becomes inhospitable. Such a species is called an **r strategist** or an opportunist species. Its extreme case is a fugitive species which is constantly being wiped out in the places it colonizes and which survives only through its ability to disperse and find new areas at a high rate. In such a population, genotypes which give a high r will be constantly favoured.

At the other end of the spectrum are individuals or species which live in long-lasting habitats such as old climax forests where environmental conditions are not so variable. The population may be very near K (the carrying capacity for the environment) and so a high r would be a disadvantage, pushing the population 'over the top'. Competitive ability to cope with crowded conditions is more important than a high reproductive rate, and these species are known as **K strategists** or stable species. Plants which are taller than their competitors or which have a bigger root system will have a selective advantage. In animals there may be selection for territorial behaviour or for dominance hierarchies, both of which allow only certain members of the population to breed.

So r **selection** favours high rates of population increase and occurs particularly in species which exploit short-lived environments, whereas **K selection** favours superiority in stable, predictable environments in which rapid population growth is unwanted. Some of the characteristics of r and K selection may be seen in figure 5.12, page 165.

r selection, with its emphasis on an individual's high reproduction rate and so its high contribution to the gene pool, is an important part of the classical idea of natural selection, whereas theories with explicit reference to K selection were not formalized until MacArthur did so in 1962.

Sexual selection

Darwin introduced the idea of sexual selection in 1871 in his book *The Descent of Man and Selection in Relation to Sex*. Sexual selection may be **intersexual** (based on choices made between courting partners) or **intrasexual** (based on competition between members of one sex, usually male, to acquire partners, and to ensure a satisfactory reproductive outcome).

1 Intersexual selection

In many species this depends on female choice. Males are often not very discriminating in their choice of female, particularly if they are able to mate frequently. Spermatozoa, being so much smaller than most ova, may be produced in enormous numbers, and the male's investment in the offspring may be restricted, in essence, to his genes. Females, on the other hand, invest more material in their gametes and may need to invest materials and energy in maternal care after the offspring are born. In many animal species, particularly in insects, the female may acquire enough spermatozoa in a single mating to last her lifetime, and she is discriminating about which male she will accept. Sometimes her choice is frequency-dependent and depends on the relative numbers of different male phenotypes, as we have already seen in the case of the Scarlet Tiger Moth.

2 Intrasexual selection

This results from competition between males, or occasionally between females, to be able to reproduce satisfactorily. The competition may act before mating in many different ways. Some males or females may be more efficient than others in actually finding mates. In species where territories are established or where there is a **dominance hierarchy** certain males will be more likely to mate. They may have acquired a good territory or have defeated other males in contests for females. The contests may be very real and dangerous, or may be much more ritualized and so be unlikely to cause real injury. Sometimes the structures which have a selective advantage in the context of courtship, whether to attract the female or to fight other males, may be a disadvantage at other times of life. A large head of antlers on a stag may enable it to win its contests with other males and so allow it to mate with more females. However, the large antlers are energetically expensive to grow, and may make the bearer more obvious, and so more vulnerable, to predators (or even to men who are out stag-shooting and who want to mount the head and antlers on a wall).

Competition may also act after mating has been accomplished. In mice the presence of a different male can, by means of a **pheromone** he produces, prevent the uterine implantation of blastocysts produced by a previous mating. This means that the previous mating fails, and the new male will father the next batch of offspring. In some species such as locusts a successful male will prevent the female from mating with others by remaining attached to her for a long period both before and after copulation. Even if, as in many animal species, the attachment is not physical, a male may guard his chosen female or females and try to prevent the close approach of other males. This is sometimes achieved by continually rounding-up his mated harem, but it can also result from the movement of a mated pair away from the vicinity of unmated, wandering males.

The effect of sexual selection, whether it is intersexual or intrasexual, is to produce a satisfactory reproductive outcome, and the mating systems which have evolved increase the probability of such a satisfactory outcome occurring. Particular mating systems have evolved to give optimal results in particular ecological and behavioural conditions.

Sometimes it pays either sex to have more than one mate (**polygyny** if a male mates with many females as in the red deer or the lion with their harems; **polyandry** if a female mates with many males, as in the queen honey bee on her mating flights) but sometimes **monogamy** is more useful. Which is the best strategy depends on the species. If, as in some birds, the father is needed to provide parental care both before the offspring are born by helping to incubate the eggs and afterwards by bringing food to them, then a monogamous system is of benefit. In mammals the female is tied to the offspring in their early days because she provides the milk to feed them, and so the male is not specifically needed to bring them food. Whether he stays with the family or leaves the female shortly after mating depends on whether he can contribute to the survival of his offspring in other ways. If the mammal is herbivorous he can provide food neither for the offspring nor the mother, and he tends to be polygynous and an absent father. A carnivorous male can bring food for the female while she is feeding and guarding the young and, although he may be polygynous and have a number of females living in the same social group, he can still provide a great deal of parental care for his offspring.

In some vertebrate species a female which lays many eggs may be so exhausted by egg production that she cannot give any parental care to her offspring. The male, having provided the spermatozoa, may then safeguard his investment by looking after the developing eggs and young, as happens in the Three-spined Stickleback. In large birds such as geese and swans the male can actively chase most potential predators away from the nest or young, whereas this is not feasible for the smaller males of duck species. Ducks rely on the cryptic colouration of the female and the concealment of the nest, and the presence of a brightly coloured male would be a disadvantage. The totally male parental care of the Stickleback, the shared parental care of geese and swans, and the totally female parental care shown by ducks are all evolved strategies which give the optimal results in the particular conditions and are what Maynard Smith has called **evolutionary stable strategies**.

Group selection and kin selection
In all of the preceding types of selection we have considered the advantages and disadvantages of a particular trait to an individual, and the way in which gene frequencies may alter in successive generations of a population because of the different reproductive success rates of the individuals which compose it.

Sometimes particular traits *seem* to be disadvantageous to individuals but advantageous to the group in which they live. Imagine a species which is divided into separate, discrete sub-groups, some of which contain in their gene pools a gene which sacrifices an individual for the benefit of the group. Groups with the gene will survive better than those without, and gradually the non-sacrificial groups will be replaced by sacrificial groups. In many species of flocking birds individuals give a cry if they see a predator, and this is a signal for the flock to scatter. It seems that the signaller makes itself more obvious, and therefore more vulnerable, for the common good. Any such behavioural act which seems to decrease an individual's reproductive success for the apparent benefit of others – an *altruistic act* – appears to run contrary to the ideas of 'normal' natural selection. One could argue that if there is a gene for altruism it would soon be eliminated from the population because of the relative lack of offspring of the altruist.

So how do populations of flocking birds continue to contain altruistic individuals? There must be some benefit to the individual which more than offsets the apparent disadvantage of making the cry. It is possible that the cry could send the whole flock into random movement which could confuse the predator and make it less likely to catch any bird, including the crier. Some random bird is very likely to be caught if no bird cries, and so there is a bonus from the cry which applies to all the birds, the crier included. It might seem to be sensible for the bird which first sees the predator to slope off quietly to some safe place, leaving the other birds to be at risk. However, its very movement out of the flock, 'breaking step' as it were, might make it stand out and so be much more vulnerable than if it had stayed to take its chance with the others. So the cry, which seems at first purely to benefit the group at the expense of the individual which makes it, can be explained in terms of individual and not group selection and the group effect is an extra bonus.

There are other examples of altruistic acts which can also be explained on the basis of individual advantage. It is the genes which are maintained in the population and passed on to the next generation which are important in evolution. Each individual can be considered as the receptacle of the genes, but the important thing is that the genes are conserved and passed on. There are various ways whereby this **gene conservation** may occur. Each organism made by sexual reproduction will inherit half the genes of its mother and half the genes of its father. This means that parents and offspring will have half their genes in common but so also, on average, will siblings. There may be certain circumstances where it is better for the conservation of the genes for individuals to be sacrificed if close relatives carrying the same genes may thereby be saved.

Large and powerful males in herd animals may protect females and young at some increased risk to themselves. Parent birds may divert the attention of predators by means of distraction displays such as injury feigning. In each case the animal may conserve its genes by saving its young. The 'value' of the possible self-sacrifice of a parent to conserve genes in its young is partly determined by the parent's potential for further reproduction and also by the chance of survival of the parentless offspring. Old parents near the end of their reproductive lives have more to gain and less to lose from taking increased risks.

The evolution of such behaviour patterns has been named **kin selection** by Maynard Smith because they favour the survival of close relatives of the animal which shows them. Again, although it may look as if there is selection in favour of a whole group, the selection is really working at an individual level, conserving particular genes.

As you must already have realized, a large amount of present day thought in the evolutionary field is applied to comparisons of the costs and benefits of theoretically different evolutionary strategies and thereby to explain why a particular mode of life, or some adaptation, or a particular ecological niche, or a certain behaviour pattern, has superseded another in the evolutionary process. In the 1950s von Neumann and Morgenstern formulated **Game Theory** to analyse the optimal strategies to pursue in human conflict and Maynard Smith has extended these ideas and applied them to evolutionary problems by means of simple models.

One such model deals with an imaginary species in which the members may have one of two strategies during the conflicts which arise when they are trying to find mates. An individual may be a 'hawk' or a 'dove'. Fights between 'hawks' are likely to be fierce affairs which increase in ferocity until one contestant is seriously injured and loses. Contests between 'doves' are ritualized, neither contestant being injured and the beaten contestant finally withdrawing because of exhaustion. A population made up entirely of 'hawks' would rapidly become extinct as they would kill each other. A population made up entirely of 'doves' would survive well. Both the winner and the loser in a 'dove'/'dove' contest would spend time and energy in the contest; the loser, in addition, would lose the chance to reproduce. However, as each 'dove' is likely to win about half its contests, each would manage to reproduce at some time. But consider what would happen if a 'hawk' arose in such a 'dove' population. It would do extremely well at first, winning all its fights as its 'dove' opponents would very rapidly withdraw. The number of 'hawks' would therefore increase in subsequent populations, and so also would the chance of 'hawk'/'hawk' confrontations. A stable population would be part 'hawk', part 'dove'; each individual would then do reasonably well. 'Hawks' would meet 'doves' sufficiently often to offset their loss through 'hawk'/'hawk' encounters. 'Doves', although they would invariably lose to 'hawks', would avoid injury by their rapid withdrawal and so would survive to meet other 'doves'. So the apparently best result 'for the good of the species' – a population made up entirely of 'doves' – does not evolve because selection acts on individuals and not on the whole group. Of course what happens in the real world is likely to be much more complicated than in the model; a model such as this, however, forms a useful basis for discussing the possible profit or loss which could result from particular behaviour patterns.

In this chapter we have been looking at the different ways in which **systematic changes** in the **gene frequencies** of **populations** (the process of evolution) can be brought about. Such changes may be maintained and continued in populations in such a way, and to such an extent, that speciation occurs and this we consider in the next chapter.

SUMMARY

The Hardy-Weinberg formula can only apply to an idealized population in which gene frequency change does not occur. Natural populations are likely to change in gene frequency in succeeding generations. The changes may result from many causes. Sometimes **meiosis** may be affected in such a way that a heterozygote produces more gametes containing one of its alleles than those containing the alternative allele. Gametes may also differ in their capacity to survive long enough to meet another gamete for fertilization. **Mutations** may occur at any time to be passed on to the next generation. A natural population is not closed as required by the Hardy-Weinberg formula but may be increased in size by the **arrival of immigrants** or decreased by the **emigration** of some of its members. Both could affect gene frequencies. Even within a natural population which appears to be closed, **mating may not be at random** but occur more often between particular members.

However, the most important cause of gene frequency change within a population results from the fact that some organisms are **better adapted to survive** within the particular environment. Surviving better gives them an increased chance of reproducing and thereby passing on a selection of their genes to their offspring. The better adapted organisms leave **more surviving offspring** and so the gene frequencies within subsequent generations will change by means of this **natural selection**. There are **many types** of natural selection which result from environmental variation. Alleles which might be strongly selected in one area or at one time may be a disadvantage in other areas or under the conditions which apply at a different time. So **natural selection is a constantly changing force** which may lead populations first in one direction of gene frequency change and then in another.

FURTHER READING

Allen, J A, Clarke, B C, 'Evidence for apostatic selection in wild passerines', *Nature*, vol. 220 (1968) pp. 501–502.

Bennett, D, Bruck, R, Dunn, L C, Kylde, B, Shutsky, F, Smith, L J, 'Persistence of an introduced lethal in a feral house mouse population', *American Naturalist*, vol. 101 (1967) pp. 538–539.

Bennett, D, Dunn, L C, Arntz, K, 'Genetic changes in mutations at the *Tt* locus in the mouse', *Genetics*, vol. 83 (1976) pp. 361–372.

Bishop, J A, Hartley, D J, Partridge, G G, 'The population dynamics of genetically determined resistance to Warfarin in *Rattus norvegicus* from Mid Wales', *Heredity*, vol. 39(3) (1978) pp. 389–398.

Bodmer, W F, 'The genetics of homostyly in populations of *Primula vulgaris*', *Phil. Trans. Roy. Soc. B.*, vol. 242 (1960) pp. 517–549.

Bodmer, W F, Cavalli-Sforza, L L, *Genetics, Evolution and Man* (W.H. Freeman and Co., 1976).

Cain, A J, Sheppard, P M, 'Selection in the polymorphic land snail, *Cepaea nemoralis*(L)', *Heredity*, vol. 4 (1950) pp. 275–294.

Clarke, C A, Sheppard, P M., Thornton, I W B, 'The genetics of the mimetic butterfly, *Papilio memnon*', *Phil. Trans. Roy. Soc. B.*, vol. 254 (1968) pp. 37–89.

Clarke, C A, Sheppard, P M, 'Further studies of the genetics of the mimetic butterfly, *Papilio memnon*', *Proc. Roy. Soc. London. B.*, vol. 184 (1971) pp. 1–14.

Connolly, K, 'The genetics of behaviour', in Foss, Brian M, (ed.), *'New Horizons in Psychology I'* (Penguin Books, 1966) pp. 185–208.

Cook, L M, *Coefficients of Natural Selection* (Hutchinson and Co. Ltd, 1971).

Cook, L M, Wood, R J, 'Genetic effects of pollutants', *Biologist*, vol. 23 (1976) pp. 129–139.

Cox, E C, Gibson, T C, 'Selection for high mutation rates in chemostats', *Genetics*, vol. 77 (1974) pp. 169–184.

Darlington, C D, 'The evolution of polymorphic systems', in Creed R, (ed.), *Ecological Genetics and Evolution* (Blackwell, 1971).

Dunn, L C, 'Variations in the transmission ratios of alleles through eggs and sperm in *Mus musculus*', *American Naturalist*, vol. 94 (1960) pp. 385–393.

Friedmann, Milton J, Trager, William, 'The biochemistry of resistance to malaria', *Scientific American* (March 1981).

Gibb, J, 'The breeding biology of the Great and Blue Titmice', *Ibis*, vol. 92 (1950) p. 507.

Greaves, J H, Rennison, B D, 'Population aspects of Warfarin resistance in the brown rat, *Rattus norvegicus*', *Mammal Rev.*, vol. 3 (1973) pp. 27–29.

Hickey, W A, Craig, G P, 'Genetic distortion of sex ratio in a mosquito, *Aedes aegypti*', *Genetics*, vol. 53 (1966) pp. 1177–1196.

Jones, J S, 'Evolutionary genetics of snails', *Nature*, vol. 285, no. 5763 (29 May 1980) pp. 283–284.

Kerney, M P, Cameron, R A D, *A Field Guide to the Land Snails of Britain and North West Europe* (Collins, 1979).

Kimura, M, 'Genetic variability maintained in a finite population due to mutational production of neutral and nearly neutral alleles', *Genet. Res.*, vol. 11 (1968) pp. 247–269.

Kimura, M, Ohta, T, 'Protein polymorphism as a phase of molecular evolution', *Nature*, vol. 229 (1971) pp. 467–469.

Krebs, C J, Myers, J H, 'Population cycles in rodents', *Scientific American* (June 1974).

MacArthur, R H, 'Some generalised theorems of natural selection', *Proc. Nat. Acad. Sci. USA*, vol. 48 (11) (1962) pp. 1893–1897.

Mather, Kenneth, 'The genetical structure of populations', *Symp. Soc. Ex. Biol.*, vol. 7 (1953) pp. 66–95.

Mather, Kenneth, *Human Diversity* (Oliver and Boyd, 1964).

Maynard Smith, John, 'The evolution of behaviour', *Scientific American* (September 1978).

Maynard Smith, John, *The Evolution of Sex* (Cambridge University Press, 1978).

Nicholson, A J, 'An outline of the dynamics of animal populations', *Australian Journal of Zoology*, vol. 2(1) (1953) pp. 9–65.

O'Donald, P, 'A further analysis of Bumpus' data: the intensity of natural selection', *Evolution*, vol. 27 (1973) pp. 398–404.

Peacock, W J, Miklos, G L G, Gabor, George L, 'Meiotic drive in *Drosophila*: new interpretations of the segregator distorter and sex chromosome systems', *Adv. Genet.*, vol. 17 (1973) pp. 361–409.

Pianka, E R, 'On r- and K-selection', *American Naturalist*, vol. 104(940) (1970) pp. 592–597.

Popham, E J, 'The variation in the colour of certain species of Arctocorisa (Hemiptera, Corixidae) and its significance', *Proc. Zool. Soc. London*, vol. 111 (1941) pp. 135–172.

Popham, E J, 'Further experimental studies on the selective action of predators', *Proc. Zool. Soc. London*, vol. 112 (1942) pp. 105–117.

Thoday, J M, Boam, T B, 'Effects of disruptive selection. II. Polymorphism and divergence without isolation', *Heredity*, vol. 13 (1959) pp. 205–219.

Thoday, J M, Gibson, J B, 'The probability of isolation by disruptive selection', *American Naturalist*, vol. 104 (1970) pp. 219–230.

Wickler, W, *Mimicry in Plants and Animals* (Weidenfeld and Nicolson, 1968).

The results of evolution

THE SPECIES CONCEPT

Any observer or investigator of life on earth must be struck by its diversity. Each organism can be classified into the particular and discontinuous sets which we call species. Its position is determined not just because it happens to look very like other organisms in the species – there are many examples of organisms which can mimic others which are fundamentally rather different from themselves – but mainly because it can breed successfully with the other members of the species. If a species is defined as a group of organisms which can actually or potentially interbreed and which therefore share a common gene pool, and which are reproductively isolated from all other groups, then there may be great problems in allocating some organisms to particular species.

Such a definition can only apply to sexually reproducing organisms. Some organisms, although fundamentally asexual, regularly exchange genetic material as in bacterial recombination, and so can be accepted under this definition. Others may be purely asexual and yet they may be designated as species even without the breeding qualification. There may also be other problems. Often the organism must be named from a limited number of specimens, which may be recently dead or even in fossil form. Organisms within a species may be highly variable showing differences for sex or age, or various polymorphisms, or even pathological differences, and so newly discovered organisms may be difficult to classify accurately.

However, the **species concept** is a useful and necessary way of dealing with the enormous numbers of different types of organisms (even if a pragmatic modern view is that a species is just a category into which a taxonomist will group individuals). The construction of a modern classification with its hierarchy of kingdom, phylum, class, order, family, genus and species, may take into account two different kinds of events – the order of descent from common ancestors, and the extent of divergence from these common ancestors. If we accept that evolution has occurred then present day species must have been derived by a series of changes from earlier species. This could have occurred in two main ways. In **phyletic** evolution an entire ancestral population could have retained its identity and its internal breeding structure, but have changed sufficiently over a long period of time eventually to warrant a new name. In **speciation** part of an ancestral population split off in some way, the original and the sub-populations then evolving (as a result of the various processes we discussed in chapter 5) in different directions.

The order in which the sub-populations split from the ancestral population, and the extent of divergence from the ancestors form the basis of a very lively debate on possible methods of classification. Imagine three species arising at the same time from a common ancestor. Two of these species subsequently develop only minor differences; the third diverges markedly from the other two because it has a major evolutionary innovation. If the three species are classified purely on the basis of their order of descent, then they would all be equally related. If, on the other hand, the extent of divergence is used as a basis for classification, then species three would be clearly separated from the other two. The great debate on classification has emerged from the consideration of how much weight should be given to each view.

METHODS OF CLASSIFICATION

There are three main schools of thought.

Phenetics

Pheneticists, such as Peter Sneath and Robert Sokal, feel that it is impossible to classify organisms objectively according to genealogy (tracing ancestral patterns). Instead they investigate morphological (phenotypic) similarities and differences between organisms. To construct phenetic classifications sophisticated mathematical tools are needed and so this type of classification (now often called **numerical** taxonomy) has become more important as computers have been developed. A large number of characters are measured in the different organisms being studied and **coefficients of overall similarity** are calculated. The organisms are then placed in pairs with the highest coefficients of similarity; the pairs formed are then compared with each other and so on.

One of the main claims of pheneticists is that this method of classification eliminates the need for making subjective judgements. The characters chosen are measurable mathematically and so can be dealt with objectively. However, the first three words of the previous sentence show that the problem of subjectivity still remains. Which characters are chosen? Are all characters regarded as being equally important for deciding similarities and differences? The answers to these questions depend on subjective judgements.

Phenetics, depending as it does on morphological similarities, may also group together organisms which show convergent evolution rather than close ancestral relationships.

Cladistics

The principles involved in this method of classification were formulated in the 1960s by Willi Hennig, a German entomologist who has recently died, and supported by Niles Eldredge and Joel Cracraft. The term comes from the Greek *klados* (a branch or young shoot) and the original principles laid emphasis on the branching sequence in which species arise. It is assumed that all species present at one time are descended from ancestral forms, and that no species is ancestral to another living at the same time. A belief in evolution is implicit in cladistics. As in phenetic systems organisms are grouped together according to the number of characters they share. Such characters must show homology, however, and not be the result of convergence. There are striking similarities between the eyes of cephalopod molluscs and vertebrates, for example; the similarities, however, do not show recent shared ancestry but convergent evolution in two very different groups of organisms. The more homologous characters shared by two species, the more closely they are related. Some characters, such as multicellularity or the presence of a backbone, are found in so many organisms that these shared **primitive characters** are only of value in the initial classification into very wide groups. However, organisms have acquired other characters during the course of evolution. By identifying these

derived characters and seeing how many organisms share them it is possible to draw branching trees or cladograms. The four animals, domestic cat, lion, seal and man, for example, share many primitive and derived characters which have led to them being classified together as mammals. Man, however, has certain derived characters, such as a large brain and the very good binocular vision which is possible because of his flattened face, which are not found in the others. The first branch point in a cladogram therefore separates man. The lion and the domestic cat have more characters in common than either has with the seal, and so the second branch point separates the seal. Finally the lion and the domestic cat are separated by differences between them.

A problem may arise for the cladist in distinguishing between primitive and derived characters, but there are techniques which help to

Figure 6.1   Different types of classification

A   Cladistics

The figure shows a cladogram of four mammals. Man has certain derived characters like a large brain and binocular vision which are not shared with the other three mammals and so the first branch point separates him. The second branch point separates the seal because the lion and domestic cat have many shared derived characters.

B   Phenetics

The construction of a tree from phenetic data

Five species A, B, C, D and E are measured and their phenetic similarities recorded in the matrix of step 1. Similarity is measured on a scale ranging from 0 (maximum dissimilarity) to 1 (complete identity). The tree is constructed by searching the matrix of step 1 for basic pairs – pairs which have the maximal similarity value. For example, the 0.95 similarity between A and B is higher than between A and any other species, or between B and any other species. So A and B form a basic pair. C and D can also be seen to form a basic pair. The basic pairs are then joined at their recorded level of similarity (at the right of the table). Each basic pair is then considered as a single unit in step 2, and the matrix is recalculated. Each similarity value in step 2 is an average of all the individual species values. The 0.525 for AB with CD is the average of 4 values: A with C(0.65); A with D(0.6); B with C(0.45); B with D(0.4). Again basic pairs are searched for – this time there is only 1 – the 0.525 value for AB with CD. AB and CD are then joined at this level in the tree, and the process repeated in step 3. The matrix here has only 1 entry, the average of A with E, B with E, C with E, D with E, and this gives a value of 0.3. The phenetic tree is now complete.

solve it. One of the most important is 'out-group comparison'. If the characters which are being considered within a group of species are also found in species outside the group, then they are more likely to be primitive than derived.

When the ideas underlying cladistics were first formulated, a great polarization developed amongst taxonomists between the cladists on

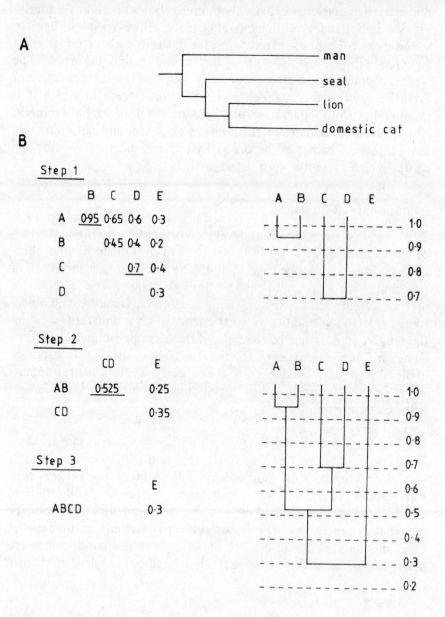

the one hand and the evolutionary systematists (see the next section) on the other. A further polarization has now occurred with the development of **transformed cladistics** as an offshoot of the original cladistics. Its main proponents are Colin Patterson at the British Museum (Natural History) and Gareth Nelson and Norman Platnik at the American Museum of Natural History. Transformed cladistics continues to use **classical cladistic techniques**, but without any underlying assumption that evolution has taken place. However, it is difficult to see how the concept of shared primitive or derived characters may be accepted without an assumption that evolution has occurred. The system of transformed cladistics is now seen by many as a rejection of evolution itself rather than as a classification system in which it is claimed that evolutionary ideas play no part. But the theory of evolution is supported by many different kinds of evidence, and not merely by any evidence which may be derived from classification systems, no matter which system is considered.

Evolutionary systematics

This form of classification was developed principally by Ernst Mayr. A combination of **genealogy** and the extent of overall similarities allow **evolutionary trees** to be constructed. The trees take account both of branching sequence (and hence timing) and of the amount of divergence between the species being considered. There are no objective rules to follow and the wide experience of taxonomists in assessing the relative importance of character differences or similarities is fundamental. As with classical cladistics, the basis of the classification lies in the concept of 'relatedness' between species. Evolutionary systematists see this in terms of 'genes in common'; classical cladists think of 'characters in common'.

Each of these standpoints may have advantages and disadvantages. There is probably no *best* way of classifying organisms. Pheneticists, with their emphasis on morphology, may confuse homologies with convergence. Cladists look for branching sequence but tend to ignore the amount of divergence between the two branches. Evolutionary systematics depends rather more on subjective judgements than do the other methods, although even in those some subjective decisions are needed. Sometimes they all agree about a classification but often there is a difference of opinion (which may generate a lot of intellectual heat!) between them.

Consider the evolution of reptiles and birds. The ancestral group which gave rise to the crocodiles and birds included the dinosaurs, but the common ancestor of *all* the reptiles (including the snakes, lizards and turtles) and the birds was a much more distant reptile group. This late splitting-off of the birds means that birds and crocodiles *could* have more characters in common than the crocodile shares with other present day reptiles. However, after divergence, the birds evolved much more rapidly than the reptiles and have many distinctive features such as wings, feathers and homoiothermy. Since birds have strongly diverged from reptiles, evolutionary systematists place them in a separate class Aves, leaving all the reptiles, crocodiles included, in the class Reptilia. The same type of classification, based not on branching but on morphological similarities and differences, would be given by pheneticists. The cladistic school, on the other hand, would construct a cladogram in which birds and crocodiles were in one section and all the other reptiles were in another.

Accepting that there will continue to be discussion and controversy about the way in which organisms may be classified, let us now go back to thinking about what might occur during the splitting of an ancestral population to make what might eventually become a new species. The new sub-population must accumulate enough genetic differences to prevent interbreeding with the original population and so reproductive isolation is of very great importance. Without some form of reproductive isolation any genetic variations will continue to play their part in the total gene pool for the population.

TYPES OF ISOLATION

Isolation may be prezygotic and exclude the formation of zygotes by preventing the meeting of gametes from the old and the new populations. If zygotes are made, their particular sets of mixed genes may be prevented from playing their expected part as parents of future generations by various means, and so postzygotic isolation may occur.

Figure 6.2 lists some of the ways in which isolation may be brought about, which we shall now consider in more detail.

A Premating isolation

1 Potential mates do not meet:
    (a) geographical isolation;
    (b) ecological or habitat isolation;
    (c) seasonal or temporal isolation.
2 Potential mates may meet but do not mate:
    (a) changed courtship behaviour;
    (b) changed signals.
3 Mechanical difficulties may prevent the gametes coming together.

B Postmating isolation

1 Gamete mortality
2 Hybrid mortality or reduced hybrid viability
3 Hybrid sterility
4 Hybrid breakdown

Figure 6.2 Isolating mechanisms *Based on table in Mayr.*

Premating isolation

Potential mates do not meet

Here isolation may be obvious if the two populations are in different **geographical** areas. This may happen as a result of a change in the course of a river isolating non-swimmers on the two opposite banks, or from a rise in sea level producing a chain of islands from an original strip of land. It may also result from the migration of part of a population to a different geographical area. However, organisms may occupy the same geographical area but still be effectively isolated from each other during their breeding time. Sometimes they occupy **different habitats** within a region. Fish may spawn in different parts of a river depending on whether the river bed is gravel, mud or sand, or according to the water flow. Mosquitoes may frequent different water areas – *Anopheles labranchis* and *Anopheles atroparvus* live near brackish water whereas *Anopheles macullipennis* live near running fresh water.

Organisms may share the same geographic area and the same habitat and still be reproductively isolated from each other if their **breeding seasons** do not overlap. Plants may produce their pollen at a time when the flowers of different plants have not opened or when the stigmas are not receptive. Animals may reach sexual maturity at different times.

Potential mates meet but mating does not occur
Ethological (behavioural) isolation results from a weakening or absence of sexual attraction between males and females of the different sub-populations. It may be that there is some change in the **courtship behaviour** carried out by either sex, or there may be changes in **visual**, **auditory** or **chemical signals**. A variation in the breeding plumage of a bird, or a slight variation in its song might be enough to bring about assortative mating. A change in the chemical structure of a pheromone may mean that it is no longer effective as a sex attractant.

Mechanical differences may prevent the gametes from coming together
Flowering plants, depending as they do on some outside vector to transfer pollen from one plant to another, have evolved along particular pathways which maximize their chances of pollination. This has often involved co-evolution between the plant and a particular insect which may be attracted and then carry out the pollination manoeuvres in a highly specialized way. Such specialization means that there is likely to be little pollination occurring between different species. In animals differences in shape of the genitalia may make the insemination of the female very unlikely even if copulation is attempted.

Postmating isolation

Gamete mortality
Sometimes the first stage of reproduction, the transfer of gametes from one organism to another, may be accomplished but the transferred gametes may then die. Pollen grains may not be able to germinate on the 'wrong' stigma; pollen tubes from those grains which have managed to germinate may not grow far enough through the style to reach the ovules in the ovary. Spermatozoa may be inviable in the genital ducts of the 'wrong' female animals. Patterson found an insemination immune response to the wrong sperm in female *Drosophila* which led to the swelling of the vaginal walls and the death of the sperm.

Hybrid mortality or reduced hybrid viability
Hybrids which arise may be eliminated before they reach sexual maturity, and therefore make no contribution to future gene pools. Sometimes the hybrids die very soon after they are formed. Cross-fertilization is possible between goats and sheep but the hybrid embryos

die very early. In some organisms hybrids may survive and even be fully fertile, but reduced viability means that they are unlikely to survive long enough to reach sexual maturity and to breed successfully. In ducks, fully fertile hybrids between Mallards, *Anas platyrhynchos*, and Pintails, *Anas acuta*, may be produced under laboratory conditions. The two species do not normally interbreed in nature, however, even though they may live and nest in the same area. The hybrid offspring are less viable than their parents, possibly because they do not fit the ecological niche of either parental species, and so any hybrids which may be produced in the wild are unlikely to last long enough to reproduce.

Hybrid sterility

Sterility of a hybrid will ensure that it cannot make a contribution, by means of its gametes, to the ancestry of future generations. The hybrid may be somatically vigorous – a mule produced by crossing a male donkey, *Equus asinus*, with a female horse, *Equus caballus*, is capable of working over longer periods and in more stringent conditions than either of its parents. Its sterility results from its inheritance of a haploid set of thirty-two chromosomes from its horse mother and a haploid set of thirty-one different chromosomes from its donkey father, producing an impossibility of homologous matching in meiosis. The same type of matching-up problem arises if triploid organisms are formed from the crossing of a diploid with a tetraploid, and so triploid offspring are sterile.

Hybrid breakdown

In this case the hybrids may be fully fertile and capable of producing an F_2 generation, but *this* generation is of greatly reduced viability. The cotton species *Gossypium barbadense*, *Gossypium hirsutum* and *Gossypium tomentosum* give vigorous and fertile F_1 hybrids, but the F_2 plants either die in the seed or seedling stage, or they survive to produce very poor plants.

TYPES OF SPECIATION

Genetic isolation is of fundamental importance in the process of speciation, and the various forms of genetic isolation mean that there are various forms of speciation.

Allopatric speciation

Allopatric means 'in another place' and so this is speciation because of geographic isolation. As we have said, the daughter populations may arise from all sorts of geographic causes but once they are isolated they may evolve in different directions. Small emigrant populations of any sort may show all the rapid changes which are associated with limited size such as significant random genetic drift and the effect of the founder principle. They are also protected against gene flow from the parental population. The environment which they inhabit may subject them to intense selection pressures which are different from those affecting the parental population.

Parapatric speciation

This may occur if a daughter population, although occupying a different geographical area, remains in contact with its parent population along the borders. Intense selection pressures in the two areas, which lead the two populations on to different evolutionary paths, may be sufficient to overcome gene flow in either direction at the border, and genetic differences will accumulate within the two populations. Clines may be good candidates for parapatric speciation if selection is strong and operates in different directions at each end, and if gene flow along the cline is weak.

Sympatric speciation

This occurs within an existing population – 'in the same place' – and so can only arise if the bearers of mutations either preferentially mate with each other, or if postzygotic isolation removes any hybrids which may be formed. Sometimes large changes such as polyploidy in plants may immediately isolate the bearer from others. It would be of no evolutionary significance if the plant was thereby useless as a producer of the next generation. However, many plants are capable of vegetative reproduction and so the polyploid plants could be maintained and their numbers increased to a level where they could interbreed. If the polyploids were of even number sets of chromosomes they would have no meiotic homologue matching-up problems, but any cross breeding between diploid and tetraploid plants would produce sterile triploids.

So isolation would be produced between the diploids and the tetraploids. Triploids could become fertile and form a different species if their chromosome sets were doubled, in a process called allopolyploidy, to give hexaploids.

The modern bread wheat, *Triticum aestivum*, is descended from three wild grasses – *Aegilops speltoides*, *Triticum monococcum boeticum* and *Aegilops squarrosa*, each of which was diploid (2n = 14). Hybridization followed by allopolyploidy occurred between *Aegilops speltoides* and *Triticum monococcum boeticum* to give another wild wheat, *Triticum turgidum dicoccoides* (2n = 28), a cultivated form of which is the macaroni wheat *Triticum turgidum durum*. Further hybridization and allopolyploidy between *Triticum turgidum dicoccoides* and *Aegilops squarrosa* produced the hexaploid *Triticum aestivum* (2n = 42).

Polyploidy is rare in animal species because of their lack of asexual reproduction which could maintain and increase the polyploid strain to a point where sexual interbreeding becomes feasible. However, there are other ways in which a particular section of a population may rapidly become isolated. Many animal parasites are host-specific. In 1864 a sub-population of the Hawthorn fly, *Rhagoletis pomonella*, started to lay its eggs under the skin of apples and so the original and the sub-population were isolated by this change in ecological niche. As more than 50 000 species of insects are host-specific, such a mode of isolation and speciation might be important.

Most examples of sympatric speciation, however, will require a long period of time for the necessary genetic differences to accumulate. Selection must overcome gene flow between one part of the population and another. If hybrids are of lower fitness than the other members of the two diverging populations, then individuals which cross breed to form hybrids will leave fewer offspring than those which do not.

Figure 6.3   Polyploidy and the history of wheat

A   *Aegilops speltoides*
B   *Triticum monococcum boeticum*
C   *Aegilops squarrosa*
D   *Triticum turgidum durum*
E   *Triticum aestivum*
F   *Triticum turgidum dicoccoides*
For explanation see text.

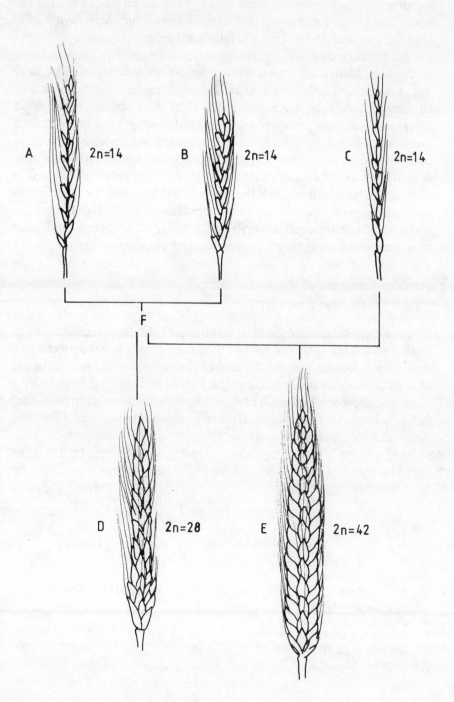

Genotypes which reduce the possibility of cross-fertilization with the other form will be at an advantage. Selection in their favour will gradually reduce and eventually eliminate cross-matings.

This has been demonstrated in laboratory experiments carried out by Koopman. *Drosophila pseudoobscura* and *Drosophila persimilis* only rarely form hybrids in the wild, but will readily hybridize under laboratory conditions. Koopman kept the two species together in laboratory cages and removed all the hybrid offspring produced in each generation. At the beginning of the experiment somewhere between 22 per cent and 50 per cent of all the offspring produced in each cage were hybrid, but after six generations the number of hybrids had dropped to 5 per cent. Continuation of the experiment brought the number of hybrids down to 1 per cent in some populations. An isolating mechanism seemed to have evolved very rapidly under these special conditions in which selection against hybrids was total.

STAGES OF SPECIATION

The postzygotic isolation which results from the elimination, sterility or F_2 breakdown of any hybrids which may be produced between two sympatric populations is obviously more wasteful of biological resources (because the hybrids are on a 'dead end' path) than the forms of prezygotic isolation which prevent the hybrids from being produced at all. Selection will therefore tend to promote prezygotic rather than postzygotic isolating mechanisms.

So speciation is likely to proceed in two stages. To begin with, some reproductive isolation occurs as an incidental byproduct of genetic

---

Figure 6.4 Premating isolation in Australian Tree Frogs

A   Recording sites and geographic ranges of *Hyla ewingi* and *Hyla verreauxi* in south-eastern Australia.

B   oscillograms of the mating calls of *Hyla ewingi* and *Hyla verreauxi* from allopatric and sympatric populations.

a   allopatry
b   sympatry (western)
c   sympatry (eastern)

Taken from Littlejohn.

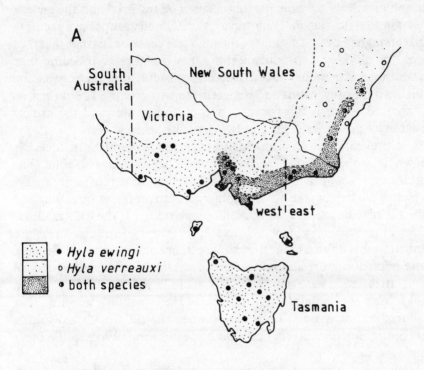

A

South Australia | New South Wales

Victoria

west | east

::: • *Hyla ewingi*
::: ◦ *Hyla verreauxi*
■ • both species

Tasmania

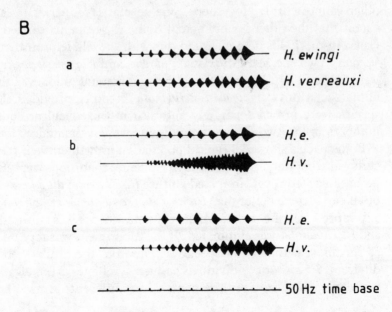

B

a — *H. ewingi*
    — *H. verreauxi*

b — *H. e.*
    — *H. v.*

c — *H. e.*
    — *H. v.*

— 50 Hz time base

diversity between two populations. If the loss of fitness in the hybrids is not too great it is possible for this stage to be reversed and the genetic isolation to be lost again. If, however, the isolated groups have accumulated enough genetic differences through one cause or another that they would be unlikely to produce fertile offspring in cross-breeding then natural selection would favour the less wasteful prezygotic isolation which would prevent mating between the two groups. From then on differences between the two groups would increase, and the groups could exist in the same area as two separate species.

The prevention of hybridization in sympatric populations is clearly shown in two species of Tree Frogs from south-eastern Australia. The species, *Hyla ewingi* and *Hyla verreauxi*, overlap for part of their range.

Male Tree Frogs, like other amphibians, attract mates by calling and the calls may be studied using oscilloscope traces. If the calls made by males of the two species which live in different geographic areas are compared it is obvious that they show rather similar patterns. Because these populations are allopatric there is no chance of the wrong female being attracted. However, such calls made in the overlapping region could lead to wasteful hybridization. Here natural selection has favoured an increasing difference between the calls, with a resulting prezygotic isolation. In this case both species can survive in the overlapping area.

Sometimes, however, two species which are rather similar to each other cannot survive together – one species is 'superior' to such an extent that the other is eliminated. Some experiments carried out by Gause in the 1930s demonstrated this with the ciliate protozoan *Paramecium*. Two species, *Paramecium aurelia* and *Paramecium caudatum*, were cultured in tubes in which a constant amount of food, in the form of the bacterium *Pseudomonas aeruginosa*, could be provided. The two species were grown separately and also in mixed culture and their numbers in the two different conditions were estimated.

Both species showed a typical population growth curve if they were cultured separately, and so the environmental conditions were shown to be suitable. However, in mixed culture *Paramecium aurelia* seemed to be able to survive better than *Paramecium caudatum* – *P. aurelia* gradually approached the population level found when it was cultured alone, and *P. caudatum* was eliminated. If the experiment was repeated using *P. caudatum* with another species, *P. bursaria*, the results were rather different. *P. bursaria* individuals contain symbiotic green algae and so their physiological requirements are different from those of

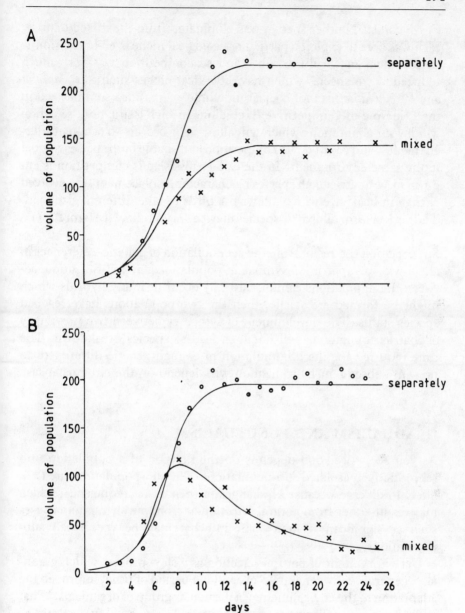

Figure 6.5   Competition between species of *Paramecium*

A *Paramecium aurelia*, separately and in mixed culture.
B *Paramecium caudatum*, separately and in mixed culture.
Both species can survive well if cultured separately, but only *Paramecium aurelia* managed to survive in mixed culture.   *Taken from Gause.*

P. caudatum. Neither species was eliminated from the mixed culture; each was able to exploit a particular ecological niche even in the simple environment of a culture tube, and so competition was very much reduced. Two species which have identical niches such as *P. aurelia* and *P. caudatum* cannot coexist indefinitely – an idea summarized in the **Principle of Competitive Exclusion**. In such a situation, selection will favour adaptations which will allow their bearers to occupy different niches, and so cut down competition and enable both species to exist in the same environment. In the Galapagos Islands competition is cut down in sympatric finch species by character displacement, the evolved change in beak size which allows the birds to take different size food. The beak size for allopatric species may be very similar (see figure 2.7).

So speciation can result from the accumulation of genetic differences in allopatric, parapatric or sympatric populations. If hybridization between the populations cannot naturally occur or if any hybrids which might be formed are sterile, then the two populations have reached species status. The development of sympatry between two previously allopatric populations will not then lead to species breakdown. The time required for the accumulation of enough genetic differences to prevent subsequent hybridization will depend on the circumstances.

GRADUALISM AND PUNCTUALISM

As you see, speciation depends on the division of a population into reproductively isolated sub-populations. The sub-populations are then likely to diverge because of changes in their gene frequencies, which may result either from natural selection or from purely random causes such as significant genetic drift in what may be very small sub-populations.

Darwin thought of natural selection as a slow process which gradually, generation by generation over a long period of time, improved the adaptation to the environment of a particular group of organisms. 'That natural selection will always act with extreme slowness, I fully admit' he wrote in *The Origin of Species*. But the two concepts of gradualism and natural selection are not necessarily tied together as the two sides of the same coin. One may be accepted without the other. Lamarckists, for example, might well argue for gradualism during evolution by the inheritance of acquired characters, without accepting natural selection.

Selectionists do not have to subscribe to the idea of gradualism. Although Darwin strongly supported both concepts in his theory of evolution, there were, nevertheless, some problems which he found difficult to solve. The fossil record which was available to him did not seem to support a gradualistic view (see pages 26–29). There were great gaps in the record (which, it was hoped, would eventually be filled by more investigation) but also (and of more importance) there was frequently a sudden appearance of many new forms at the same geological time. Evolution seemed to be proceeding in 'fits and starts'. Long periods of time in which certain species were present in a relatively unchanged form were interspersed by periods in which other species suddenly appeared, thereafter to remain relatively unchanged, in their turn, for a long time. The long periods of stasis were punctuated by short periods of change. This episodic, rather than gradual, change was named punctualism or punctuated equilibrium.

So how is it that evolutionary change sometimes appears to be gradual and sometimes episodic? The most general prediction which can be made from the theory of natural selection is that episodes of rapid evolution should coincide with periods when the direction of selection is changing. Certainly this is seen in the recent changes in Peppered Moth populations as the amount of industrial pollution has fallen. The rapid evolution of insecticide resistance in insect pest populations exposed to a new insecticide is similar. Although the rate of evolution is high, however, the changes within these populations occur in a gradual way. Such changes within populations and species have been named **micro-evolution**.

The time required for speciation to be accomplished is very variable. If speciation is occurring in small, isolated populations, whether they are a long way from the parent population as in oceanic islands, or are small isolates on the periphery of the main population, then evolution may be very rapid. The isolated populations may be faced by a harsh environment in which selection may be powerful and leading in a different evolutionary direction from that faced by the main population. The gene pool of the isolated population is likely to be very restricted, and the small size of the population might allow random genetic drift to become important.

A new species may be formed in a few hundred or a thousand years – a time which in geological terms is like the blinking of an eye. Fossils from any particular geological time would be likely to be of the parent species but not of the new incipient species. It would be surprising if any fossil evidence could reveal the intermediate stages of a process which, in geological time scales, is equivalent to the laying down of an extremely thin stratum. The large size of the parent population or species would mean that change within it would probably be slow and therefore its fossils would remain unchanged over a long time span. Within the species small changes could be selected to fit the changing environment. Whether or not these adaptive changes could be made would depend partly on the rate of change of the environment and partly on the alleles available in the gene pool. The genetic resources of a species are finite. Eventually there may not be alleles of the right type or at the right frequency to continue to match the changing environment, and the species will become extinct. A peripheral isolate could then take over and it, in its turn, would become large. The initially rapid genetic changes would slow down as its increasing size would lead to genetic inertia. If species turn-over of this type occurs within each lineage, with species becoming extinct and other species taking their place for a time until they also become extinct, then we would expect the fossil record to show the episodic changes which are, in fact, found.

It has been argued that gradual, even if rapid, changes within isolated populations cannot always account for the large anatomical changes which may occur during **macro-evolution** (the name given to changes above species level). These may require much larger genetic re-arrangements than point mutations – changes such as chromosomal inversions or translocations, or the changes in chromosome number in polyploidy. Most of the changes will fail in that the organism which bears them dies. However, occasionally one might survive as what Richard Goldschmidt called a 'hopeful monster' in his book *The Material Basis of Evolution* in 1940.

The punctuated equilibrium theory of macro-evolution sees the initial source of variation as being random relative to the course of evolution, but also sees this variation as involving relatively large changes. However, even though the initial variation may be produced in a random fashion, it is possible that certain trends may occur during the

Figure 6.6   The production of evolutionary trends by means of the differential origin and survival of species

There is a gradual trend in one direction because more speciation events occur in that direction, and because species with these morphologies last longer.

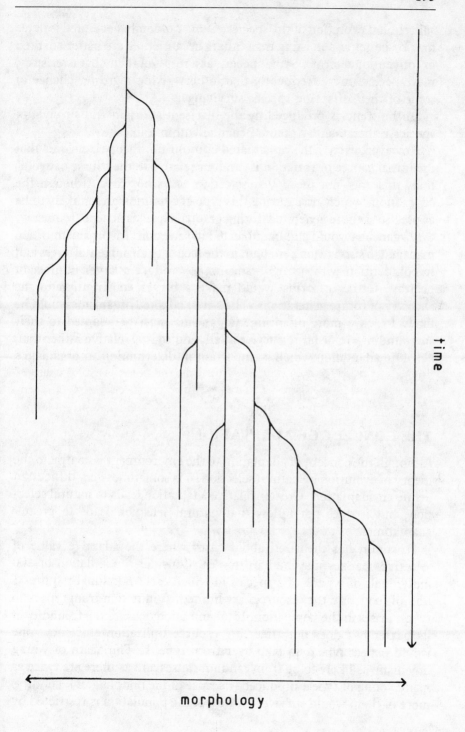

time

morphology

subsequent evolution of the species. Some randomly generated isolates may be better at persisting than others because they are better adapted in conventional terms. Some species, as a result of their characteristics, may speciate more frequently than others giving a greater chance of some of their offspring species surviving.

So the trend is produced by the differential origin and survival of species rather than by gradual change within a lineage.

However, even if the punctuated equilibrium theory considers that the initial source of variation is random relative to the course of evolution, this does not mean that selection plays no part. Although the adaptations which arise during this type of evolution are unlikely to be as precise as those generated during gradual changes in gene frequency, the organisms would still be subjected to selection. An organism which managed to survive long enough, in the face of environmental stress and in competition with other organisms, to produce a larger number of healthy, fertile offspring would make a bigger contribution to the ancestry of future generations. The better adapted organisms would be likely to leave more offspring. Organisms must be adapted to their surroundings in order to survive at all, and so it is relative rather than absolute adaptation which is important in determining an organism's fitness.

THE CONCEPT OF ADAPTATION

Although the concept of adaptation to the environment is central to the theory of evolution by natural selection, it is sometimes very difficult to define an adaptation. Greater relative adaptation leads to natural selection, but it does not follow that natural selection leads to greater adaptation.

Lewontin has theorized about cases where the adaptive value of selective changes may be hard to see. Consider a population of 100 individuals of type A of a species in which each individual requires 1 unit of food, and the resources are limited. A mutation from type A to type a arises in the population. If the mutation doubles the fecundity of its bearer but does not affect its resource utilization efficiency, one would expect type a to take over from type A. The death of young organisms is likely to be from random causes and as there are twice as many young of type a produced (because of the doubling of fecundity) more of them would survive. However, the population is restricted by the limited resources and so population size will stay at 100 individuals. If the mutation has no effect on fecundity but doubles the efficiency at food utilization, again type a will take over from type A. As each unit of food will feed two type a but only one type A individual, the population will eventually consist of 200 type a. But then it will stop growing. Both of these final populations result from selection but it is difficult to say which would be better adapted. Increased fecundity would help to buffer a population against accidents, but the offspring would be vulnerable to predators which take immature forms. However, individuals with increased resource utilization efficiency would be more vulnerable to predators which take adults in a density-dependent way. The two theoretical populations may be equally well adapted to face the hazards they meet, although they have taken two different paths.

Natural selection results from variations in the relative adaptation shown by organisms to the particular set of environmental pressures to which they are subjected. Each aspect of the organism's physiology, morphology or behaviour can be seen as a solution to a particular environmental problem. We can either start from the problem and search for the answer, or else start from an answer and then try to find the problem it solved. For example, the problem of a shortage of available water for a plant may be solved in various ways. A larger root system would allow some plants to gain more of the water and therefore to survive better. Alternatively protection against water loss by means of a thick cuticle, or by having stomata in grooves or pits or surrounded by hairs, would enable these plants to survive on a lower water uptake.

The problem posed by the necessity for species members to recognize each other for mating and so to avoid the biological wastage of producing infertile hybrids may be answered by the evolution of distinctive courtship patterns or the production of different pheromones.

It may be more difficult to define the problem if all we have is an answer, as may occur in fossil species. *Stegosaurus* was a dinosaur, about six metres in length, which lived in the late Jurassic period. Fossils suggest that it had a series of flattened, roughly triangular plates arranged in a double alternating row down the entire length of the neck and trunk, and almost to the end of the tail. The tip of the tail had two pairs of long spikes. The plates had thickened bases and it is likely that they were attached to the skeleton by tough ligaments.

Stegosaurus, like many other dinosaurs, had short front legs and much longer back legs, but it appears to have walked on four legs rather than with the bipedal gait shown by many of the carnivorous dinosaurs.

Figure 6.7   Stegosaurus

Stegosaurus, a reptile from the late Jurassic period, had an alternating row of triangular plates extending down its back. The plates might have been used for gaining or losing heat, or for defence.

Its skull was very small and long with a single row of about two dozen small teeth in each half jaw, and the size of its brain cavity suggests that its brain was very tiny. *Stegosaurus*, could have been vulnerable to attack by larger, carnivorous dinosaurs. So the large plates could be the adaptive, protective answer to the problem of attack. However, there is no indication of any armour over the rest of the body, and it seems that *Stegosaurus* could have been easily crippled by a side attack. The plates could have made the animal look larger and so have been important in defence, but in view of its relatively small size this is unlikely to have given it much, if any, protection. It is possible that the plates could have been used for recognition in courtship, but their number and size would suggest a very wasteful use of biological resources. Such large flat extensions of the body could have been used for the gathering or dissipation of heat. It is not known whether *Stegosaurus* (or any other dinosaur) was **endothermic** or **ectothermic** but the plates could have been a useful temperature regulating device in either case. The plates had a porous internal structure which suggests that they had a good blood supply. An endotherm could lose heat when necessary from such large surfaces particularly by increasing the blood flow to the plates.

Alternatively, an ectotherm could, by standing so that the plates received as much of the sun's radiation as possible, warm up the blood passing through them. So there are many possible problems which may have been solved by the evolution of the plates and although temperature regulation seems the most likely we cannot be certain that this is correct.

Even with living organisms it is often difficult to decide what type of adaptation might be an improved answer to an environmental problem. We might not fully understand the nature of the problem, and there might be an interplay of environmental factors which we cannot evaluate. One could imagine that an increase in the length of the legs of a herbivorous grazing animal would enable it to run faster. This would allow it to escape from predators more easily, enabling a longer-legged animal to leave more offspring than the more easily captured shorter-legged members of the population. But the longer legs need more biological resources to grow and maintain and may reduce the feeding efficiency of the animal, as well as being more breakable. Longer-legged animals may stand out from the background and so be more likely to be seen and chased by predators. Even if the longer legs do produce the predicted higher running speed this may be irrelevant if predators consistently take old animals which are at the end of their reproductive lives. It is also likely, of course, that during the time when the prey may be evolving to possess longer leg bones the predators will not be at an evolutionary 'stand-still'. The problem the predators need to solve is to catch enough prey animals to fuel their own reproduction, and certain adaptations may allow them to do this more efficiently. Both the predators and the prey may evolve to solve problems which are themselves being altered.

Organisms are continually evolving to fit an environment which is constantly changing. The paths taken by the organisms depend on their starting points – they can only use the genetic resources already present in the gene pool – although other paths may theoretically be more efficient. It is a bit like a stranger in town asking a local for directions to a particular place and being told that it would be better not to start from here. Previous evolution will determine the starting points for new evolutionary paths and the number of possible adaptations will be limited.

So populations are continually adapting to the environment and, as the environment is itself changing, they are continually lagging behind. Each species must evolve as fast as it can in order to survive, and natural selection operates mainly to enable organisms to maintain a state of adaptation rather than to improve it. Leigh Van Valen of the University of Chicago has summarized the idea as the **Red Queen Hypothesis** – an allusion to Lewis Carroll's 'looking glass land' where, as Alice discovers from her encounter with the Red Queen, constant running is needed in order to stay in the same place.

If the hypothesis is correct, one would expect to find differences in the methods of reproduction found in species which inhabit environments which produce different challenges. A species living in a tropical environment where competition from other species may be very strong would need to retain a large and varied gene pool, and high recombination rates in sexual reproduction, in order for it to keep pace. Where competition is lower, as may happen in temperate and sub-arctic environments, there would not be the same need for a high potential for change and **parthenogenesis** or various forms of self-fertilizing hermaphroditism would be expected. The strength of competition may also be higher in the centre of a species range where the environment may be suitable for many species than round the periphery where numbers in general are lower and where the environment may suit some species more than others.

Many studies of plant and animal species have shown that there is a difference in the reproductive strategies which have evolved under these varied conditions. Colonizing or weedy plants are often **apomictic** or self-fertilizing, and within a plant species the frequency of self-compatibility and self-fertilization is higher in marginal than in central populations. Parthenogenetic races of animals are found at higher altitudes and at higher latitudes than their nearest sexual relatives. They are also found, as are weedy plants, in disturbed habitats. Parthenogenesis or self-fertilization would be a useful characteristic for colonizing organisms as a single individual could start a colony, and many parthenogenetic races of animals are found on islands or in island-like habitats.

A complex biological environment containing many different competitors, predators and pathogenic species could only be countered by a gene pool which is large enough and complex enough to give a great number of potential evolutionary paths, and so recombination and cross-fertilization are necessary adaptations to these conditions. Many organisms which are primarily asexual or are self-fertilizing may reproduce by sexual out-crossing at times. When this happens recombination in the hybrid offspring will give rise to many new clones or pure lines.

The 'best' of these will be selected for individual environments and will then continue along their paths of limited evolutionary change until the next set of genetic rearrangements in their next sexual out-crossing.

The individual environments may be geographically very close. Hamrick and Allard found that gene frequencies at many loci in the wild oat, *Avena barbata*, over a 183 metre transect of a Californian hillside were correlated with a decrease in humidity from the foot of the hill to the top. Even if there is a continuing chance of gene flow between populations, it is possible for selection to overcome this. The 'Park Grass' plots at the Rothamsted Experimental Station are plots in a long, thin field which have been subjected to carefully monitored fertilizer treatment since 1856, and in which various grasses are grown. Snaydon, in work reported in 1970, showed that Sweet Vernal Grass, *Anthoxanthum odoratum*, grew most rapidly in plots in which the soil had the same mineral constitution as that in which the parents had grown, and so selection for these particular mineral conditions had occurred. However, Sweet Vernal Grass, like many other grasses, is wind pollinated and so gene flow could occur between a population and its neighbours. Even though gene flow was detectable for up to one point eight metres downwind and up to forty-six centimetres upwind at a boundary, there was an abrupt change in the characters of the plants within five centimetres of the boundary.

In relatively simple biological environments organisms are likely to become genetically more similar as inbreeding or asexual reproductive patterns gradually decrease the options open from a restricted gene pool. Specific environmental differences such as a range of humidity or variations in the minerals available from the soil will allow selection to establish different populations even in the very limited areas we have described above. Clinal patterns may be established over larger geographical areas. Seed size in Burr Clover, *Medicago hispida*, decreases with increasing altitude; petal length in Storksbill, *Erodium cicutarium*, increases with increased mean rainfall; leaf length in Foxtail Barley, *Hordeum nodosum*, decreases as the mean temperature decreases. In each case the plants will have been selected for, and will be in balance with, their environments.

FUTURE DEVELOPMENTS

Man has, during the course of the history of horticulture, deliberately selected certain plants as the basis for his crop plants. His modern methods of cultivation have led to an increasing restriction of the gene pool in order to improve the productivity of his chosen plants. Environmental differences can now be eliminated. The use of controlled soils, particular light intensities and temperatures, and a particular mix of fertilizers means that certain types of crop plants can now be grown over large areas of the earth. However, the benefit of increased productivity must be measured not only against the obvious costs implicit in the horticultural methods but also against the hidden cost of the loss of the genetic diversity found among plant populations growing in natural conditions in many parts of the world. Whenever the new, more productive plants are imported into an area then the old, primitive plants are discarded. The loss of the world's genetic resources held in primitive crop plants could lead to disastrous results in the future if the conditions necessary for the growth of the highly selected cultivated varieties cannot be maintained.

The highly successful wheats produced by the Rockefeller team in Mexico and the new rice varieties developed by the International Rice Research Institute in the Philippines during the recent decades have transformed the agricultural picture over much of the world where under-nutrition was prevalent, in what has been heralded as the 'Green Revolution'. The main aim of a great improvement in crop productivity has been achieved with a subsequent reduction in the misery of hunger and starvation for vast numbers of people. However, wherever the new varieties are introduced, the centres for genetic diversity of more primitive crop plants are being eliminated. Morally it is extremely difficult to try to balance the present need to improve the nutrition of people who are already on the earth and who are undernourished against our possible future needs of introducing alleles from primitive crop plants which have been safeguarded in their natural environments.

The problem of the continuing loss of genetic resources has, however, been recognized and certain guidelines about their maintenance by means of regional and international gene banks have been suggested by the Food and Agriculture Organization of the United Nations. Although it sounds relatively straightforward to set up gene banks to store genetic resources, there are enormous practical problems to be solved. The genotypes cannot be stored for long periods of time as seeds, but the populations of plants which may be periodically grown to produce the next batch of seeds for storage will be subject to selection in the unnatural conditions of a laboratory or a cultivated plot. The collections of plants would probably be small populations and so significant genetic drift could occur, leading to the fixation of deleterious alleles. It might also prove to be very difficult to keep the plants isolated from others, and so out-crossing may occur. However the recognition of the problem, and the continued search for original, primitive varieties which can be safeguarded in the field should at least reduce the rate of loss of genetic resources.

Man is, as far as we can judge, the only animal who can deliberately alter the pattern of evolution of other organisms. Up until the last decade his power of being able to change the characteristics of his chosen animals and plants has depended on selection of those organisms which best suited his needs at the time. Such artificial selection is partially offset by the chance events during meiosis and fertilization which result in offspring not being replicas of their parents. Although the chance nature of selection in sexually reproducing animals offers a great and enjoyable challenge to the large numbers of people involved in trying to breed the 'perfect' wire-haired Fox terrier, or Plymouth Rock cockerel, or any of the other animals bred for show, it has disadvantages in the breeding of farm stock. Continued inbreeding will reduce the chance element as more and more of the genes become homozygous, but such continued inbreeding may lead to the fixation of deleterious alleles. The initial encouraging rate of improvement in the desired characteristic will diminish as inbreeding continues and, because of the accumulation of deleterious alleles during the selection process, fecundity and viability may also decrease. I M Lerner showed in a survey of poultry flocks which were selected for their ability to lay large eggs that the eggs which were more likely to hatch were those which were smaller than the mean. In populations of *Drosophila melanogaster*, in which selection for high numbers of thoracic bristles was being carried out, there is a loss of many selected lines because of sterility. In general, if a population is subjected to selection for a particular trait then those organisms which show the trait most strongly have lower reproductive fitness than the average members of the whole population. The harmonious interaction of the particular mix of alleles which has evolved by natural selection is disrupted by artificial selection and eventually no further selection for the trait is possible. The breeder may then need to introduce new alleles by out-breeding.

Artificial selection is therefore a useful, but not perfect, way for a breeder to change the characteristics of his chosen stock. The problems of selection are different with those plants which reproduce asexually. Particular clones may be selected but the rate at which large numbers of desired plants may be obtained depends on the form of vegetative reproduction. Potato plants, for example, produce many tubers during a season's growth whereas Daffodils produce few new bulbs. However, the techniques of **micropropagation** which are now available and which are constantly being improved allow breeders to produce new plants at a high rate from a very small number of stock plants with the characteristics they desire. Very small pieces of plant material may be cultured in sterile containers and will rapidly produce plantlets if given the correct proportion of hormonal and nutritional factors in the medium. The process can be applied to plants which normally reproduce wholly by sexual means, as well as to species which reproduce asexually.

Recently the possibility of cloning animals by artificial means has been brought into public interest, partly because of the techniques such as *in vitro* fertilization which are now available in human reproduction (test tube babies) and partly because of a great fascination with science fiction 'predictions' for life in the future. However, the ideas behind cloning are not new. Early in this century Loeb working with sea urchins and Spemann working with salamanders showed that nuclei taken from cells of very early embryos were capable of organizing the development of whole, normal embryos if transplanted into cells from which the nuclei had been removed. More recent cloning experiments have used amphibians such as the Leopard Frog, *Rana pipiens*, and the South African Clawed Toad, *Xenopus laevis*. Some work has also been carried out on insects and fish. Mammals present greater practical problems in cloning experiments but success may produce great agricultural advantages. Instead of the chance nature of selective sexual crossing, animals could be cloned for outstanding milk production, or for a better quality of wool, or for any other characteristic which is shown by a single animal. So the breeder could achieve a herd of animals which are all genetically similar in a very short time. However, it is important to remember that, although the animals would be genetically similar, it is unlikely that phenotypically they would be carbon copies of each other. Heritability in the chosen trait might be relatively low (see figure 4.4) and so differences in the environment would result in different phenotypes being produced.

The rapidly developing techniques whereby one may introduce new genes into organisms by means of genetic engineering suggest that enormously advantageous innovations in crop plants may be possible within the relatively near future. The incorporation of the gene, normally found in various species of bacteria, for nitrogen fixation would drastically reduce the need for expensive nitrogenous fertilizers. In human lives the possible excision of deleterious alleles such as those causing sickle cell anaemia and their replacement by alleles coding for normal haemoglobin by means of genetic manipulation is, as yet, a dream for the future. However, the implantation of human genes into bacteria has already achieved great success in the commercial manufacture of important large molecules such as insulin.

It is only about thirty years since Watson and Crick published their important paper on the structure of DNA and yet work in the new fields of genetic engineering and other forms of biotechnology may enable man to determine the future course of the evolution of his own, and of many other species. What is particularly important is that man, with all the advantages that his own evolution of consciousness and conceptual thought have given him, must not destroy the environment in which other organisms may exist. All species eventually disappear as a result of natural changes in their environment. We must not accelerate the process either by over-exploitation or by making the environment so different from that to which the species has been adapted by a long process of natural selection that the organisms can no longer survive.

SUMMARY

Organisms, like everything else, must be given names by which they can be identified. Obviously this does not mean a different name for each individual organism, like Rover or Scamper for a pet dog, but a name which applies to all dogs. Such a name is a **species name**. All members of one species are theoretically able to **interbreed** and so share a **common gene pool**, and any **offspring** they produce are **fertile**. However, a different definition must be given for **asexual organisms** or those which constantly show **self-fertilization**. Here the extent of **similarity** between individuals can be used as a basis for a decision. So although the **species concept** is a necessary and useful idea, there is **no one definition** of a species which can apply to all organisms. Groups of related species are placed together in a **genus**. The domestic dog, *Canis familiaris*, shares a genus with the wolf, *Canis lupus*.

Given that all organisms can be placed in a species and a genus, it becomes less confusing if related genera are grouped into **families,** families into **orders,** orders into **classes,** classes into **phyla,** and phyla finally into **kingdoms.** However, there may be much argument among **taxonomists** (experts in classification) about deciding on the 'right' place in the hierarchy for any particular species. Some taxonomists, the **pheneticists,** base their argument wholly on similarities and differences between species. **Cladists** rely on the assumed pattern of the branching of species from their ancestral forms. **Evolutionary systematists** use a combination of ideas based both on the amount of similarity and on 'family trees'.

If a species consists of interbreeding individuals which share a common gene pool then there must be **reproductive isolation** between one species and another. Otherwise the gene pool would be common to both and so they would form just one species. Reproductive isolation may be produced in many ways. Individuals may not be able to mate together for **geographical** or **ecological** reasons or because their **reproductive seasons** do not match. Particular patterns of **reproductive behaviour** or **colouring** may mean that they attract mates from their own species only. Even if mating occurs between species and gametes are transferred, no hybrids may be produced because the **gametes or the zygotes die.** Sometimes hybrids may survive but as they are **sterile** they produce no offspring.

Reproductive isolation means that species can **evolve separately** from each other and the differences between them will increase. The first step needed, therefore, in the evolution of a new species is the **reproductive isolation of part of a population.** The **rate** at which the isolated section reaches species status varies according to its **size,** its store of **genetic variability** and the **selective effects** of its particular environment. A very small size allows alleles to be **lost by random genetic drift** because of the luck of random mating amongst few individuals. The gene pool of the isolated section is **unlikely to contain all the alleles** found in the main population and the environment which the isolated section has to cope with may make certain alleles more, or less, advantageous to their carriers. In such circumstances new species become established so rapidly that evolution seems to be proceeding in **leaps and bounds, with static equilibrium stages** in between when there seems to be little change. In other circumstances **slow, gradual change** is found.

With his increased knowledge and understanding of genetics, man is now in a position to change, by means of various kinds of biotechnology, the future evolutionary paths of many organisms including, possibly, his own future course.

FURTHER READING

Charlesworth, Brian, 'Neo-Darwinism – the plain truth', in Cherfas, Jeremy, (ed.), *New Scientist Guide: Darwin – Up to Date* (1982).

Cox, Barry, 'Seeing the wood for the phylogenetic trees', *Nature*, vol. 285, no. 5760 (8 May 1980) p. 119.

Eldredge, N, Cracraft, J, *Phylogenetic Patterns and the Evolutionary Process. Method and Theory in Comparative Biology* (Columbia University Press, New York, 1980).

Feldman, Moshe, Sears, Ernest R, 'The wild gene resources of wheat', *Scientific American* (January 1981).

Frankel, O H, Bennett, E, (eds.) 'Genetic resources in plants – their exploration and conservation', (International Biological Programme) *I.B.P. Handbook No. 11* (Blackwell Scientific Publications, 1970).

Gause, G F, *The Struggle for Existence* (Williams and Williams, Baltimore, 1934. Reprinted Hafner, New York, 1964).

Glesener, R R, Tilman, D, 'Sexuality and the components of environmental uncertainty: clues from geographic parthenogenesis in terrestrial animals', *American Naturalist*, vol. 112 (1978) pp. 659–671.

Goldschmidt, Richard, *The Material Basis of Evolution* (Yale University Press, 1940).

Goldschmidt, Richard, 'Evolution as viewed by one geneticist', *American Scientist*, vol. 40 (1952) pp. 84–123.

Gould, Stephen Jay, 'The return of hopeful monsters?', *Natural History*, vol. 86 (1977) p. 30.

Hamrick, J L, Allard, R W, 'Microgeographical variation in allozyme frequencies in *Avena barbata*', *Proc. Nat. Acad. Sci. USA*, vol. 69 (1972) pp. 2100–2104.

Hennig, W, *Phylogenetic Systematics* (republished by University of Illinois Press, 1979).

Koopman, K F, 'Natural selection for reproductive isolation between *Drosophila pseudoobscura* and *Drosophila persimilis*', *Evolution*, vol. 4 (1950) pp. 135–148.

Levin, D A, 'Pest pressure and recombination systems in plants', *American Naturalist*, vol. 109 (1975) pp. 437–451.

Lewontin, Richard C, 'Adaptation', *Scientific American* (September 1978).

Littlejohn, M J, 'Premating isolation in the *Hyla ewingi* complex (Anura:Hylidae)', *Evolution*, vol. 19 (1965) pp. 234–243.

McKinnell, R G, *Cloning: a Biologist Reports* (University of Minnesota Press, 1979).

Patterson, J T, Stone, W S, *Evolution in the genus Drosophila* (Macmillan, New York, 1952).

Ridley, Mark, 'Can classification do without evolution?', *New Scientist*, vol. 100, no. 1386 (1 December 1983) pp. 647–651.

Snaydon, R W, 'Rapid population differentiation in a mosaic environment. I. The response of *Anthoxanthum odoratum* populations to soils', *Evolution*, vol. 24 (1970) pp. 257–269.

Sneath, P H A, Sokal, R R, *Numerical Taxonomy* (W H Freeman, San Francisco, 1973).

Stanley, Steven M, *Macro-evolution. Pattern and Process* (W H Freeman and Co., 1979).

Stephens, S G, 'The internal mechanisms of speciation in *Gossypium*', *Bot. Rev.*, vol. 16 (1950) pp. 115–149.

Swaminathan, M S, 'Significance of polyploidy in the origin of species and species groups', in Frankel, O H, Bennett, E, (eds.), *Genetic Resources in Plants – their exploration and conservation*, (International Biological Programme) *I.B.P. Handbook number 11* (Blackwell Scientific Publications, 1970).

van Valen, Leigh, 'The Red Queen', *American Naturalist*, vol. 111 (1977) pp. 808–810.

Wallace, Bruce, Srb, Adrian M, *Adaptation* (2nd ed.) (Prentice-Hall Foundations of Modern Biology Series, 1964).

White, Michael J D, *Modes of Speciation* (W H Freeman and Co., San Francisco, 1978).

The application of the Hardy-Weinberg Law

The Hardy-Weinberg Law states that if p and q are the frequencies of the dominant and recessive alleles of a gene respectively so that $p + q = 1$, then the genotype frequencies may be found by calculating p^2 (dominant homozygotes), $2pq$ (heterozygotes) and q^2 (recessive homozygotes).

As the dominant homozygotes and the heterozygotes share the same phenotype, the calculations must start from the recessive homozygotes as these are the only organisms whose genotype is obvious. Some questions state the number of recessive homozygotes – see worked example 1. Other questions give the number of organisms showing the dominant character. In answering these you must remember to start by finding the number of recessive homozygotes by subtraction – see worked example 2. Although these days many students immediately turn to their calculators for even the simplest bit of arithmetic, most of the Hardy-Weinberg calculations which are set are mathematically very straightforward. It may be easier in some cases to work with decimal figures; in others it helps to leave the figures as fractions and then cancel.

Worked examples

1   The human genetic disease phenylketonuria is caused by the lack of an enzyme which can convert the amino acid phenylalanine into tyrosine (see page 78). Homozygous recessives (pp) cannot make the enzyme and are found in the proportion of 1 in 10 000 of the population. Calculate the allelic frequencies and the numbers of dominant homozygotes and heterozygotes in a population of 10 000.

1 in 10 000 are affected and so are homozygous recessive. This is q^2

$$q = \sqrt{\frac{1}{10\ 000}} = \frac{1}{100} \text{ or } 0.01$$

As $p + q = 1$, $p = \dfrac{99}{100}$ or 0.99

The frequencies and numbers of genotypes can now be found.

Dominant homozygotes = p^2

Frequency of dominant homozygotes $= \dfrac{99}{100} \times \dfrac{99}{100} = \dfrac{9801}{10\ 000}$

Number of dominant homozygotes $= \dfrac{9801}{10\ 000} \times 10\ 000 = 9801$

As two of the genotype numbers are now known, the heterozygote number may be found by subtraction. However, any error made in calculating the number of dominant homozygotes will be compounded, and so it is better to find 2pq and then check by addition.

Heterozygotes = 2pq

Frequency of heterozygotes $= 2 \times \dfrac{99}{100} \times \dfrac{1}{100} = \dfrac{198}{10\ 000}$

Number of heterozygotes $= \dfrac{198}{10\ 000} \times 10\ 000 = 198$

The numbers of genotypes in a population of 10 000 are 9801 dominant homozygotes, 198 heterozygotes and 1 recessive homozygote.

2   In maize, tallness results from the presence of a dominant allele. In a field of 2000 maize plants, 1920 are tall. Calculate the allelic frequencies, and the numbers of dominant homozygotes and heterozygotes in the population.

The 1920 tall maize plants are a mixture of dominant homozygotes and heterozygotes. There are 80 dwarf plants in the population (2000 − 1920) and these are homozygous recessive.

$$q^2 = \frac{80}{2000} = \frac{4}{100}$$

$$q = \sqrt{\frac{4}{100}} = \frac{2}{10} \text{ or } 0.2 \qquad \text{As } p + q = 1, p = \frac{8}{10} \text{ or } 0.8$$

The frequencies and numbers of genotypes can now be found.

$$\text{Dominant homozygotes} = p^2$$

Frequency of dominant homozygotes $= \frac{8}{10} \times \frac{8}{10} = \frac{64}{100}$

Number of dominant homozygotes $= \frac{64}{100} \times 2000 = 1280$

$$\text{Heterozygotes} = 2pq$$

Frequency of heterozygotes $= 2 \times \frac{8}{10} \times \frac{2}{10} = \frac{32}{100}$

Number of heterozygotes $= \frac{32}{100} \times 2000 = 640$

The numbers of genotypes in a population of 2000 are 1280 dominant homozygotes, 640 heterozygotes and 80 recessive homozygotes.

Practice calculations

The following set of figures show numbers of recessive homozygotes in 9 hypothetical populations. Each population has a different allelic frequency. Calculate the allelic frequencies and the numbers of dominant homozygotes and heterozygotes in each population. The answers are at the end of the appendices.

Population	Total number in population	Number of recessive homozygotes
1	600	6
2	600	150
3	300	108
4	300	243
5	500	80
6	600	294
7	600	24
8	750	480
9	900	81

Appendix B

Experiments in selection

1 The reaction of garden birds to variously coloured 'prey'

In this experiment, a known number of artificial 'grubs' are placed on the grass in an area which is usually visited by garden birds. The 'grubs' must not be harmful to the birds, and any food dye used must affect the colour but not the taste of the 'grubs'.

The 'grubs' are moulded from pastry made from a mixture of flour and lard in the proportions of 5:2 by weight. Add green food colouring with the binding water to one part, and brown food colouring with the binding water to the rest of the pastry as it is being mixed. Mould the pastry into 'grubs' which are cylindrical, and about 1 centimetre long and 0.5 centimetre in diameter.

Having made the 'grubs', you must now prepare the grass area. Cover the area systematically with a 1 metre quadrat frame, placing 2 'grubs' randomly within each square. Keep a careful count of the numbers of green and brown 'grubs' distributed. Leave for some hours, or for a day, and then count the remaining 'grubs'.

The experiment may be carried on over many days but the spatial distribution of the 'grubs' should be changed daily. In this experiment you are testing the effect of camouflage on predation. If the 'grubs' are placed on soil, rather than on a grassy background, the advantage of being green is likely to decrease. By varying the proportions of green and brown 'grubs' you can also test for frequency dependent selection. Can you devise a method which might show disruptive selection? Other food colours may be used to see whether birds avoid novel 'grubs'. Paint some 'grubs' with yellow and black stripes. Are such warning colours of some apparent value in preventing predation?

A variation of this experiment uses students instead of garden birds in the selection process.

2 An exercise in selection by students

For this experiment wooden cocktail sticks are used. Some are rolled in grass green emulsion paint and are left to dry; others are left their original natural wood colour. The sticks are then placed over a grassy area by some students, using the same distribution technique as in the previous experiment. Other students are then sent out to retrieve as many as possible within a fixed time. The proportions of green and natural sticks may be varied to see if frequency dependent selection is operating.

(Sticks remaining on the grass at the end of the experiment must be removed to protect the lawn mower!)

Answers to calculations

In all the following answers, p is given before q, and the genotypes are in the order – dominant homozygotes; heterozygotes; recessive homozygotes.

1 0.9, 0.1; 486, 108,   6
2 0.5, 0.5; 150, 300, 150
3 0.4, 0.6; 48, 144, 108
4 0.1, 0.9;   3, 54, 243
5 0.6, 0.4; 180, 240, 80
6 0.3, 0.7; 54, 252, 294
7 0.8, 0.2; 384, 192, 24
8 0.2, 0.8; 30, 240, 480
9 0.7, 0.3; 441, 378, 81

Remember that these are hypothetical populations, and the figures for recessive alleles are not those which would be expected in real populations.

Glossary

Allele One of the possible forms of a particular gene. Alleles occupy the same locus on homologous chromosomes and one allele can change into another by mutation.

Amino acid sequence The arrangement of amino acids in a protein.

Aneuploidy A cell normally contains some multiple of a whole set of chromosomes (see ploidy). Aneuploidy is a variation in chromosome number which affects only part of a set. For example, in a diploid cell there may be an extra chromosome giving three of one sort (trisomy) or a missing chromosome giving only one of that sort (monosomy).

Apomixis A form of reproduction which, although it appears to be sexual, does not involve the union of gametes.

Assortative mating Mating which is non-random.

Carcinogens Agents such as certain chemicals or certain forms of radiation which can induce cancer.

Chromosome inversions Breaks in chromosomes sometimes occur. The broken section may then rejoin to the rest the 'wrong way round' to give a chromosome inversion in which the gene order has, as a result, been altered.

Clone A collection of genetically similar organisms. Clones are produced naturally by asexual reproduction. Small clones, such as identical twins, may also be produced by the very early division of a zygote into two groups of cells, each of which develops into a separate organism.

Deoxyribose One of the types of pentose sugars which are found in nucleic acids (see nucleic acids).

Dihybrid cross A genetic cross between organisms which are heterozygous for the two different genes which are being investigated.

Dominance hierarchy A term used in animal behaviour studies to describe the way in which animals in a social group are organized into levels of 'importance'.

Dominant gene A gene which has as strong an effect when present as one allele in a heterozygote as when present as two alleles in a homozygote.

Ectothermic An organism which relies on the absorption of heat from the environment.

Endothermic An endothermic animal can maintain a constant body temperature by means of physiologically generated heat.

Eukaryote An organism which has its DNA (see nucleic acids) in chromosomes within a nucleus bounded by a nuclear membrane.

Fecundity The numbers of potential offspring of an organism.

Gene The length of DNA which codes for the formation of a particular polypeptide or protein. The fundamental unit of inheritance.

Genetic code The arrangement of bases within the DNA molecule which, when transcribed into RNA molecules, organizes the production of proteins with a particular amino acid sequence (see nucleic acids).

Genetic crossing-over The interchange of sections of homologous chromosomes during meiosis.

Genome A set of chromosomes corresponding to the haploid set of a species and therefore a set of all the different genes (see ploidy).

Genotype The genetic constitution of an organism.

Heterozygous Homologous chromosomes have two different alleles at a particular locus.

Homozygous Homologous chromosomes have identical alleles at a particular locus.

Inbreeding Breeding between close relatives.

Isotope One of a number of possible forms of a chemical element. The atoms of the isotopes of an element have the same numbers of protons and electrons (and therefore have the same chemical properties) but have different numbers of neutrons (and therefore have different masses). Radioactive isotopes are unstable and go through a series of changes emitting radiation of various types.

Linked genes Genes which are on the same chromosome.

Locus The position of a particular gene on a chromosome.

Modifier gene A gene which modifies the effects of another gene.

Mutagen An agent which can induce a gene to mutate.

Mutation A change in the form and/or the amount of the DNA in a cell which might result in a change in the proteins formed.

Niche The way of life of an organism which enables it to survive in a particular habitat.

Nucleic acids The fundamental molecules of heredity in all organisms. Each nucleic acid molecule, a macromolecule (a very large molecule), is made up of a series of smaller units called nucleotides. A nucleotide consists of a base, a sugar and a phosphate. There are two types of sugar involved – deoxyribose and ribose. Both are pentoses – they contain five carbon atoms. The two types of sugar determine the two types of nucleic acid. Deoxyribose is found in DNA (deoxyribonucleic acid) and ribose is found in RNA (ribonucleic acid). The bases are of two main types – purines and pyrimidines. Adenine and guanine are purines; thymine, uracil (which replaces thymine in RNA) and cytosine are pyrimidines. The arrangement of the bases in the DNA determines the genetic code. Messenger RNA (mRNA) molecules are then produced by the process of transcription in which each mRNA is arranged as a sequence of bases which is complementary to a certain length of DNA. These mRNA molecules move to cytoplasmic bodies called ribosomes where translation of their code may be carried out. Molecules of transfer RNA (tRNA) pick up and 'ferry' individual amino acids to the ribosomes where the amino acids are linked together in the order specified by the mRNA.

Nucleotides The units from which nucleic acids are built up.

Parthenogenesis The production of an organism from an unfertilized ovum.

Pharyngeal pouches Outgrowths from the pharynx in embryonic vertebrates which may reach through to the outside to form gill slits, or may become part of other structures.

Phenotype The inherited characteristics shown by an organism.

Pheromone A chemical signal which can pass information from one animal to another, e.g. a sex attractant.

Phylogenetic tree A suggested pattern of evolutionary descent.

Ploidy The number of sets of chromosomes present in the cells of an organism. In higher organisms there are usually two sets of chromosomes per cell, making the organisms diploid. Gametes from these diploid organisms must have only one set (haploid) so that the diploid number will be regained at fertilization. In many higher organisms, particularly higher plants, more than two sets of chromosomes are present giving a polyploid condition. Polyploids are named according to the number of chromosome sets, e.g. triploid (three sets), tetraploid (four sets), hexaploid (six sets).

Prokaryotes Organisms which have their DNA in long threads which are not enclosed within a nuclear membrane, e.g. bacteria.

Purines and pyrimidines The constituent bases of nucleic acids (see nucleic acids).

Recessive gene A gene which has phenotypic effects only if it is in the homozygous condition.

Recombinant forms New assortments of alleles on chromosomes as a result of genetic crossing-over.

Regulator gene A gene which does not code for a protein but which controls the action of a structural gene.

Ribose One of the types of pentose sugars which are found in nucleic acids (see nucleic acids).

Ribosomes The cytoplasmic bodies at which proteins are assembled from their constituent amino acids (see nucleic acids).

Segregation The separation of the two alleles of a gene into different gametes, brought about by the movement of homologous chromosomes during meiosis.

Self-replicating   A structure which is capable of organizing the production of an exact copy of itself.

Steroid nucleus   A complex organic molecule composed of three rings, each containing six carbon atoms, joined to a fourth ring which contains five carbon atoms. Many important biological molecules such as sex hormones, the hormones from the adrenal cortex and bile acids are based on steroids.

Structural gene   See gene.

Territory   An area occupied relatively exclusively by a single animal, or by a family group.

Transcription   The production of mRNA (see nucleic acids).

Translation   The production of proteins according to the code in mRNA (see nucleic acids).

Bibliography

Berry, R J, *Inheritance and Natural History* (Collins New Naturalist Series, 1977).

Burns, George W, *The Science of Genetics – an Introduction to Heredity* (Collier Macmillan Publishing Co. Inc., 1980).

Calow, Peter, *Evolutionary Principles* (Blackie, 1983).

Cherfas, Jeremy, (ed.), 'Darwin – Up To Date', *New Scientist Guide* (1982).

Darlington, P J, *Evolution for Naturalists: The Simple Principles and Complex Reality*, (Wiley Interscience, 1980).

Darwin, Charles, *On the Origin of Species by Means of Natural Selection*.

Darwin, Charles, *Journal of Researches during the Voyage of HMS Beagle*.

Dawkins, Richard, *The Selfish Gene* (Oxford University Press, 1976).

Dobzhansky, T H, *Genetics of the Evolutionary Process* (Columbia University Press, 1970).

Dobzhansky, T H, Ayala, F, Stebbins, G L, Valentine, J W, *Evolution* (W H Freeman and Co., 1977).

Farnsworth, M W, *Genetics* (Harper and Row Publishers Inc., 1978).

Fisher, R A, *The Genetical Theory of Natural Selection* (Oxford University Press, 1930).

Ford, E B, *Ecological Genetics* (3rd ed.) (Chapman and Hall, 1971).

Gardner, Eldon J, *Principles of Genetics* (5th ed.) (John Wiley and Sons Inc., 1975).

Goodenough, Ursula, *Genetics* (2nd ed.) (Holt-Saunders International Editions, 1978).

Gould, Stephen Jay, *Ever Since Darwin. Reflections in Natural History* (Penguin Books, 1980).

Haldane, J B S, *The Causes of Evolution* (Longmans, Green, 1932).

Hitching, Francis, *The Neck of the Giraffe or Where Darwin Went Wrong* (Pan Books, 1982).

Huxley, Julian, *Evolution. The Modern Synthesis* (2nd ed.) (George Allen and Unwin Ltd., 1963).

Lack, D, *The Natural Regulation of Animal Numbers* (Clarendon Press, 1954).

Luria, Salvador E, Gould, Stephen J, Singer, Sam, *A View of Life* (Benjamin/Cummings Publishing Co. Inc., 1981).

Maynard Smith, John, *The Theory of Evolution* (Penguin Books, 1972).

Maynard Smith, John, *The Evolution of Sex* (Cambridge University Press, 1978).

Mayr, Ernst, *Populations, Species and Evolution* (Harvard University Press, 1970).

Mayr, Ernst, 'Evolution', *Scientific American* (September 1978).

Patterson, Colin, *Evolution* (Routledge and Kegan Paul in association with the British Museum [Natural History] 1978).

Ridley, Mark, 'Who doubts evolution?', *New Scientist*, vol. 90, no. 1259 (25 June 1981).

Ridley, Mark, 'How to explain organic diversity', *New Scientist*, vol. 94, no. 1304 (6 May 1982).

Roberts, M B V, *Biology, a Functional Approach* (3rd ed.) (Nelson, 1982).

Salthe, Stanley N, *Evolutionary Biology* (Holt, Rinehart and Winston Inc., 1972).

Savage, Jay M, *Evolution* (3rd ed.) (Holt, Rinehart and Winston Inc., 1977).

Shorrocks, Bryan, *The Genesis of Diversity* (Hodder and Stoughton, 1978).

Stansfield, William D, *Genetics* (Schaum's Outline Series, McGraw Hill Book Co., 1969).

Wright, Sewall, 'Evolution in Mendelian populations', *Genetics*, vol. 16(2) (1931) pp. 97–159.

Young, J Z, *Introduction to the Study of Man* (Oxford University Paperback, 1976).

Index

Achromatopsia 40
Adaptation 196–200
Adaptive radiation 26, 30, 158
Adrenocorticotrophic
  hormone 63
Agrostis tenuis 71, 157
Albinism 78, 121
Alkaptonuria 77, 78
Allen's Rule 48
Allopatric speciation 185
Altruism 169
Ames Test 95–7
Amino acid sequences 62, 81,
  214
Analogous structures 50
Aneuploidy 89, 214
Anolis 117
Antibiotic resistance 104
Antifouling paints 72
Apomixis 200, 214
Archaeopteryx 50
Area effects 147
Artificial selection 64–7, 203
Asbestos danger 97
Asellus 117
Autumn crocus 66

Bacteria:
  reproduction 87
  resistance 104, 157
Bacteriophage 87, 104
Barley 201
Batesian mimicry 144, 160
Beagle voyage 11, 13, 32
Bergmann's Rule 48
Biogeographical realms 41
Blue green algae 24
Broad bean 98
Bromouracil 95

Caffeine 98
Carcinogens 94, 214
Carrying capacity 111, 112, 148,
  166
Cats:
  Siamese 115
  Manx 147
Central dogma 82
Cepaea species 146, 147
Character displacement 36, 192
Chimpanzee 55, 56
Chordates 57
Chromosome inversions
  117, 214
Chromosome mutations 89
Cladistics 177–80, 181
Classification 49, 176–81
Clawed toad 100, 204
Clean Air Act 68, 71
Clinal variation 48, 185, 201
Clone 88, 109, 214
Cloning experiments 204
Clover 201
Coevolution 135, 183
Colchicine 66
Collared dove 139
Colonizers 35, 39
Competition 162–7

Competitive exclusion 192
Conjugation 87
Constant Final Yield 164
Continental Drift 45, 46
Convergent evolution 43, 50,
  177
Cotton species 184
Cryptic colouration 161, 168
Cytochrome c 62

DDT 98, 157
Dartford warbler 138
Darwin, Charles 11–15, 26, 27,
  31, 32, 64, 110, 166, 192
Darwin's finches 32–7, 139
Death rate 112
Demes 118, 130, 133
Directed mutation 127
Domestic cattle:
  heritability 116
  hybrid vigour 156
  lethal alleles 147, 148
  repetitive DNA 100
Domestic poultry:
  heritability 116
  selection 203
Dominance hierarchy 166, 167,
  214
Drosophila:
  balanced lethals 155
  disruptive selection 158,
    159
  electrophoresis 85, 86
  eye colour 78, 79
  heritability 116
  heterozygosity 88
  hybridization 188
  insemination response
    183
  modifier genes 144
  mutation rate 99
  resistance to DDT 157
  transposable elements 104
Ducks:
  hybridization 184
  reproductive behaviour
    168

Ear ossicle evolution 51, 52
Ectocarpus 72, 157
Electromagnetic spectrum 91
Electrophoresis 85, 99
Emigration 130–2
Escherichia coli 82, 97, 99, 104,
  128
Eugenics 122
Evolutionary stable strategies
  168
Evolutionary systematics
  180, 181
Exons 101–3

Fat Hen 163
Fecundity 12, 196, 215
Field vole 132
Fossil record:
  Cambrian 'boom' 25
  dating methods 22, 23

fossilization 20
geological time scale 21
horse fossils 27–9
microfossils 23–5
Founder effect 41, 139

Galapagos Islands 31–7, 192
Game Theory 170
Gene banks 202
Gene conservation 169
Gene control 100–6
Gene/enzyme relationship
  77
Gene flow 130, 186
Gene frequency 118
Gene linkage 76, 77, 215
Gene pool 110, 166, 175, 193,
  200
Genetic code 82–3, 215
Genetic crossing over 77, 215
Genetic engineering 205
Genetic resources 202
Genome 100, 215
Germ plasm theory 84
Gloger's Rule 48
Gondwanaland 47
Gradualism 29, 192
Gray 94
Great Tit:
  clinal distribution 49
  clutch size 152
Green Revolution 202
Gulls 49
Gunflint chert 24

Habitat choice 117
Haemoglobin 79–81
Hair dyes 97
Hardy-Weinberg Law
  118–24, 209–11, 213
Hawthorn fly 186
Heritability 116, 204
Heterozygote advantage
  153–7
Heterozygosity amount 88
Hitch-hiker theory 129
Homologous structures 50
Hormone evolution 63
Housefly resistance 157
House sparrow 48, 152
Human:
  birth weight 150, 151
  development rate 55, 56
  gene flow 132
  haemoglobin 79–81
  heterozygosity 88
  mutation rate 99
  skin colour 79, 90
  taster gene 120, 121
Hybrid breakdown 184
Hybrid mortality 183
Hybrid sterility 184
Hybrid vigour 156
Hybridization 65

Inbreeding 39, 203, 215
Incompatibility 145, 162
Injury feigning 170
Insertion sequence 104

Insulin 63, 205
Introns 101–3
Ionizing radiation 90
Isolation mechanisms 181–4

Jaw evolution 51

Krill 88

Laurasia 47
Leopard frog 204
Lethal genes 147, 155
Linkage disequilibrium 144
Lithosphere 45
Locust 167
Lung fish 44, 47
Lysergic acid diethylamide 98

Macroevolution 194
Maize:
    controlling elements 103,
        127
    heterosis 156
    mutation rate 99
Malaria 153
Malate dehydrogenase 86
Marijuana 98
Marsupials 41
Mating systems 167
Meiotic drive 125
Melanism 68–71
Mendel, Gregor 15, 76, 77
Metal tolerance 71–2
Methanogens 24–5
Microevolution 193
Micropropagation 204
Migration pressure 130
Miller and Urey experiments
    58–60
Modifier genes 144, 215
Mosquitoes:
    Aedes 126
    Anopheles 117, 157, 182
Mottled Beauty Moth 69
Mouse:
    heritability 116
    irradiation 94
    mutation rate 99
    pheromone 167
    reproduction rate 112
    tailless locus 125–6
    yellow gene 147
Mule 184
Mutagens 90–8, 215
Mutation (general) 62, 78, 79,
    81, 82, 86, 87–99, 127–30, 215
Mutation pressure 128
Mutation rates 99
Mutational currency 128

Natural selection: 12, 15, 140–71
    density dependent 162–4
    directional 149, 157–8
    disruptive 149, 158–61
    frequency dependent
        161–2
    group and kin 168–70
    hard and soft 147–8
    r and K 164–6
    sexual 166–8
    stabilizing 148, 149, 150–7
NeoDarwinism 15–16
Neoteny 56
Neurospora crassa 77, 99
Neutral mutations 17, 129
Nicotine 98

Nitrogen fixation genes 205
Nitrous acid 94
Non-Darwinian evolution
    129
Non-random mating 133, 135

Oat 201
One gene/one polypeptide
    theory 77
Operon 101, 102
Ozone layer 61

Paedomorphosis 56
Pangaea 45, 46
Panspermia 5
Panthalassa 45
Paramecium 190, 191
Parapatric speciation 185
Parental care 167
Parthenogenesis 156, 200, 216
Pea inheritance 76
Pentadactyl limb 53, 54
Peppered moth 68, 117, 144,
    157, 193
Pest species 113
Pharyngeal pouches 57, 216
Phenetics 176–9
Phenylketonuria 78
Pheromone 167, 183, 197, 216
Photosynthesis 61
Phyletic evolution 176
Phylogenetic tree 62, 216
Pigeons 64, 65
Pingelap 40
Placental mammals 41
Plankton bloom 114
Plasmids 104, 157
Plate tectonics 45
Pollution effects 67, 71
Polygene systems 76
Polymorphism 68, 160
Polyploidy 66, 89, 185, 186, 216
Population cycles 113, 114
Population genetics 110
Population growth curve
    111
Pre-adaptation 50
Primrose:
    heterostyly 145–6
    polyploidy 66
Prolactin 63
Punctualism 29, 192

Rabbit 112, 115
Radiation effects 90–4
Radioactive isotopes 22, 215
Random genetic drift
    135–8, 139, 147, 193
Random walk 129
Rat:
    reproduction rate 112
    resistance 154
Ratite birds 44, 47
Red Queen Hypothesis 200
Relative fitness 140, 141, 142,
    143
Repetitive DNA 100
Reproductive strategies 200
Retinitis pigmentosa 40
Rice 202
Ring species 48, 49
Rockefeller Team 202
Roentgen 94
Rothamsted Experimental
    Station 201

Rudd 161

Salamanders:
    Axolotls 56
    cloning experiments 204
Salmonella:
    Ames Test 96
    mutation rate 99
Scarlet Tiger Moth 162, 163,
    167
Scientific method 7
Sea urchin 100, 204
Searching image 161
Seed size 150, 151
Selection coefficient 140, 141,
    142, 143
Sex evolution 26
Sickle cell allele 80, 81, 121,
    153–4, 205
Sigara distincta 161
Sodium cyclamate 98
Somatic mutations 84
Speciation:
    stages 188–92
    types 184–8
Species concept 175
Species turnover 194
Split genes 101–3
Starling 152
Stegosaurus 197–9
Steroid nucleus 63, 217
Storksbill 201
Stromatolites 23, 24
Sulphur bacteria 61
Supergene 145–6
Surtsey 39
Survival rate 140, 141, 142, 143
Swallow tail butterfly 144, 145,
    160
Sweet vernal grass 201
Swifts 152
Sympatric speciation 185
Synonymous mutations 129

Taster allele 120, 121
Territory 112, 166, 167, 217
Tethys Ocean 47
Three-spined stickleback
    168
Thyroid hormones 63
Transduction 87
Transformation 87
Transposable elements 104
Tree frogs 189, 190
Trillium 93, 94
Tristan da Cunha 40
Tunicates 57
Two-spot Ladybird 69–71

Ultraviolet radiation 90
United Nations FAO 202

Vestigial structures 54

Wallace, Alfred Russell 11, 12,
    14, 41, 44, 110
Wallace Line 44, 47
Warfarin 154
Wheat species 186, 187, 202
Wing structure 53

Xanthine dehydrogenase 86, 99
Xeroderma pigmentosum 90

Yeast 98, 104

Zwitterion 85